THE COMMAND OF LIGHT

Rowland's School of Physics and the Spectrum

Henry A. Rowland, ca. 1890. Courtesy of the Ferdinand Hamburger, Jr.
Archives of the Johns Hopkins University. Negative #7730

THE COMMAND OF • LIGHT

Rowland's School of Physics and the Spectrum

George Kean Sweetnam

American Philosophical Society
Independence Square • Philadelphia
2000

Memoirs
of the American Philosophical Society
Held at Philadelphia
For Promoting Useful Knowledge
Volume 238

ISBN: 0-87169-238-4
US ISSN: 00659738

Library of Congress Cataloging-in-Publication Data

Sweetnam, George Kean.
    The command of light : Rpowland's school of physica and the spectrum / George Kean
    Sweetnam.
        p. cm. -- (Memoirs of the American Philosophical Society Held at Philadelphia for
    Promoting Useful Knowledge, ISSN 0065-9738 ; v. 238)
    Include bibliographical references and index.
    ISBN 0-87169-238-4 (cloth)
    1. Rowland, Henry Augustus, 1848–1901. Diffraction gratings. 3. Physics—United
States—History—19th century. 4. Physicists—United States—Biography. I. Title. II.
Memoirs of the American Philosophical Society ; v. 238.

Q11P612 vol. 238
[QC16]
530'.092--dc21
[B]

                                                                    00-044181

Contents

ACKNOWLEDGMENTS   vii

ABBREVIATIONS   ix

PREFACE   xi

INTRODUCTION   xv

chapter **1** ENGINEERING A NEW DISCIPLINE   1

chapter **2** ANALYSIS MATERIALIZED   31

chapter **3** A SCHOOL OF LIGHT   61

chapter **4** THE PHYSICS OF HEAVEN AND EARTH   87

chapter **5** THE GHOSTS IN THE INSTRUMENT   115

chapter **6** ELEMENTS OF STYLE   139

chapter **7** THE PROLIFERATION OF SPECTRAL LINES   165

chapter **8** ASTRONOMICAL LIGHT IN THE WEST   187

BIBLIOGRAPHY   211

INDEX   219

Acknowledgments

I would first of all like to thank my faculty advisors at Princeton University. Norton Wise has provided insight, suggestions, criticism, and encouragement during the project. Likewise, Charles Gillispie has given valuable encouragement, advice, and discussion, essential to the evolution of this study. Other Princeton faculty giving critical input have included Nancy Nersessian, Gerald Geison, and Michael Mahoney. Fellow students who furnished helpful commentary have included Theodore Arabatzis, Ross Bassett, John Detloff, and Teresa Hopper. And Princeton graduate secretaries Kathy Baima and Peggy Reilly have provided cheerful help throughout the long process of graduate education.

Spencer Weart of the American Institute of Physics provided valuable suggestions for the formulation of the study and for archival materials to consult. At the archives of Johns Hopkins University, Joan Grattan furnished ready help in consulting the crucial Rowland Papers, and suggestions as to worthwhile resources available. At the University of Pittsburgh, Frank Zabroski helped with access to material pertaining to instrument-maker John Brashear. At the Huntington Library in San Marino, California, Ronald Brashear furnished assistance in locating relevant materials from the papers of George Ellery Hale, and other help in navigating the collections there. Roy Goodman provided help in navigating the collections at the American Philosophical Society Library in Philadelphia. I am grateful to Professor Gerald Holton and William Andrewes for enabling me to examine concave diffraction gratings kept in the collection of historical scientific instruments at Harvard University. Likewise, I record my gratitude to Deborah Jean Warner for her help in examining instruments and spectra kept in the collections of the Smithsonian Institution in Washington, DC. I thank Alison Kemmerer of the Addison Gallery at Phillips Andover Academy for allowing me to view a large scale portrait of Henry Rowland painted in the last century by Thomas Eakins.

For discussion, I am also grateful to participants in the 1993 and 1995 history of astronomy workshops at the University of Notre Dame in Indiana, and for comments from David DeVorkin in particular. Comments at a history of science workshop at Rutgers University were also helpful.

For fellowship support, I here record my gratitude to Princeton University, and to the American Philosophical Society, for the John Slater Fellowship.

Abbreviations in Notes

WSA   Walter S. Adams Papers, Carnegie Observatories Collection, Huntington Library

JSA   Joseph S. Ames Papers, Johns Hopkins University

ANQP   Archive for History of Quantum Physics.

ApJ   *Astrophysical Journal Biographical Memoirs of the National Academy of Sciences*

HC   Henry Crew Papers, Northwestern University.

HCM   Henry Crew Papers, Microfilm, American Institute of Physics.

DCG   Daniel Coit Gilman Papers, Johns Hopkins University.

GEHC   George Ellery Hale Collection, Huntington Library.

GRIN   George Ellery Hale Papers, Microfilm Edition.

TL   Theodore Lyman Papers, Harvard University.

WFM   William F. Meggers Papers, American Institute of Physics.

PWM   Paul Willard Merrill Papers, Carnegie Observatories Collection, Huntington Library.

HAR   Henry Augustus Rowland Papers, Johns Hopkins University.

PPR   *The Physical Papers of Henry Augustus Rowland.* Baltimore: The Johns Hopkins University Press, 1902.

Preface

Most students of the history of American science would agree that Henry Rowland was the single most important figure in the founding of a modern discipline of physics in the United States. The first president of Johns Hopkins, Daniel Coit Gilman, appointed him to be professor of physics in 1875, a year before the university opened. By the time of Rowland's death in 1901, he had trained a considerable proportion of those who constituted the first American generation of professional physicists. George Sweetnam's study is the only full-scale scholarly treatment of his career at Hopkins that has yet appeared. On that score alone, it is clearly an important contribution. Sweetnam sought out and consulted many surviving manuscripts, Rowland's own papers as well as those of his contemporaries. They are to be found in a dozen different collections, the most important at Hopkins itself. He read and thoroughly mastered the existing literature. Further work on Rowland and his time will have to begin with Sweetnam's monograph, which was his doctoral dissertation.

The work is not a scientific biography, however, It is both less and more than that. It is less in that Sweetnam does not attempt a complete life of Henry Rowland and restricts the account of Rowland's physics to his invention of the diffraction grating, for which indeed he is best known, and to its importance in spectroscopy. Given that limitation, Professor Norton Wise, who directed Sweetnam's thesis, has kindly consented to provide an introductory summary of Rowland's career in physics in general. Sweetnam's book, however, is more than a biography in that his interest goes beyond Rowland's own use of the grating to its role in the formation of a set of disciples, a school of physicists of which the style and identity were defined by their use of the instrument in analyzing optical phenomena.

Unhappily, this is the only work we shall have from George Sweetnam. It is very sad that he should not have been given the opportunity to live the life and do the work of which he was capable. Sweetnam concentrated in history and science at Harvard and graduated magna cum laude in 1979. After several years of experience with journalism, first with the *Miami Herald*, and then on the editorial staff of *Science Digest*, he enrolled in the Program in History of Science at

Princeton in 1982. I then taught an introductory reading seminar for entering students and was greatly impressed from the outset with Sweetnam's ready grasp of the material, with his enthusiasm for science past and present, and with his remarkable gifts as a writer. It came as a surprise, therefore, when at the end of a highly proficient year of study, he found himself unsure whether scholarship was right for him or he for it, and elected to withdraw from a graduate study, at least for the time being, in order to try his hand at the business.

Sweetnam did not, however, give up his interest in the field or his reading. With our high opinion of his qualitites, the faculty of the Program was the more gratified that in 1989 he chose to resume his studies. I had retired by then. While doing me the honor of keeping in frequent touch, then and throughout his graduate career, he worked closely at the outset with Professor Nancy Nersessian, who later left Princeton for another opportunity. To all appearances, matters went well. Sweetnam passed his generals with flying colors and qualified for the M.A. in 1990, just over a year after returning. Professor Wise became his adviser on joining the faculty in September 1991. Sweetnam formulated the problem of his thesis clearly, and the American Philosophical Society awarded Sweetnam the John Slater Fellowship for 1992-93.

Thereafter progress on the dissertation was steady, but Sweetnam's personal situation became more complex. Teaching preceptorials, as graduate students normally do for a semester or two, was not a happy experience. Sweetnam combined a fine mind with a gentle disposition and a generous nature, but he was reserved and sometimes uneasy in this manner. Though he was never hard on those around him, and on the contrary was extremely considerate, he was very hard on himself. He was at his best and most graceful with the written word.

Sweetnam's dissertation was essentially complete by January of 1995, but he was reluctant to part with it. Looking back, one can see, and indeed one sensed at the time, that he was less and less happy with his work and with himself as the months went by. He was a perfectionist, of course. We also know now that he was not in good health. He collapsed one day a few months before finishing and had to be hospitalized briefly. He nevertheless completed his final revision in the summer of 1996 and took the final public examination in September of that year. His defense of the thesis, as I wrote him afterwards, elicited as interesting a discussion among partici-pating faculty and students as I could remember.

Sweetnam was effective in discussion among people he knew. He was not at his best amid the stresses of interviews with search committees, and he never received an offer of a job. I hoped that he might elect to pursue a writing career, but he wanted to teach and felt he had invested too much of himself in preparation to change course. Back home in Manchester, Connecticut, he had open heart surgery in March of 1997, and though he recovered physically, he was unable to overcome the depression against which he evidently had been struggling hard, as we ought to have seen much earlier. He died by his own hand on 4 September 1997.

All concerned are grateful that the Committee on Publications of the American Philosophical Society has seen fit to publish his work.

Charles C. Gillispie
Princeton, May 1998

Introduction

HENRY ROWLAND AND AMERICAN PHYSICS

M. Norton Wise[1]

In 1929 Werner Heisenberg traveled and lectured widely in the "New World" on the quantum mechanics in whose recent formulation he had played such a large part. The experience left him with the distinct impression that American physicists approached their subject with attitudes quite different from his own and from many of his Northern European colleagues. In addition to their straightforwardness, warmth, and optimism, he remarked on something at once attractive and disturbing, namely, their pragmatic attitude toward the new theory. They did not seem to care too much about its revolutionary conceptual and philosophical implications. Instead, they were impressed with it as a new tool to solve problems. Heisenberg found this attitude epitomized in a young experimental physicist at the University of Chicago named Baron Hoag, with whom he went on a fishing trip to remote northern lakes. For Hoag, the difference was that "You Europeans, particulary you Germans, are inclined to treat such new ideas as matters of principle." While in America, "Basically, physicists, even the theorists among them, behave just like the engineer building a new bridge." If the old formulae do not work adequately in the new situation the engineer corrects them. "But the basic engineering principles have remained unchanged. The same seems to be true of modern physics. Perhaps you make the mistake of treating the laws of nature as absolutes, and you are therefore surprised when they have to be changed. To my mind, even the term 'natural law' is a glorification or sanctification of what is basically

[1] Program in History of Science, Department of History, Princeton University, Princeton, NJ 08544.

nothing but a practical prescription for dealing with nature in a particular domain."[2]

This engineering attitude, though not unique to American physicists, nor even universal among them, was certainly widespread when Heisenberg wrote. And it has remained alive and well, albeit in a somewhat different form, even after the center of theoretical activity shifted from Europe to America during the 1930s. It is tempting simply to ascribe it to traditional American practicality, Yankee know-how, and pragmatism. But in the case of physics a much more concrete story pertains. George Sweetnam tells an important part of that story in *The Command of Light*, which focuses on Henry Rowland's diffraction grating and on the Hopkins school of physics built up around their use. For under Rowland's leadership, from the beginning of the new institution in 1875 until his death in 1901, Hopkins became by far the most important nursery of physics in America. This preeminence is manifest in the fact that from 1879 to 1901 no less than 165 graduate students studied with Rowland at Hopkins, of whom 43 are distinguished with "stars" in *American Men of Science*. Most came for advance training without taking a degree. Of the 45 students who obtained the Ph.D. 30 are starred.[3]

Thus Rowland's physical laboratory produced a whole generation of physicists who staffed departments emerging throughout the country, and many of them carried his style of work with them. But Rowland was not working in a vacuum. Although he shared the fondness of some historians for casting him in the heroic role of a lone champion of research struggling against the odds to fulfill a vision that few could see, it would be more appropriate to regard Rowland as playing a particularly prominent part in a movement with many actors. The engineering style emerged from their interaction. Perhaps the most salient single fact in this larger picture is that all three of the most famous American physicists of the late nineteenth century—Rowland, Michelson, and Gibbs—began their careers with an engineering orientation. Rowland took his degree in engineering from Rensselaer Polytechnic Institute with a thesis on the stream engine in 1870. Albert A. Michelson honed his skill with optical devices at the Naval Academy in Annapolis, where he graduated in 1873, before embarking on the physics that would earn him the first American Nobel Prize in 1907. A new study of Michelson and some of the younger people from his Chicago

---

[2] Werner Heisenberg, *Physics and Beyond: Encounters and Conversations*, trans. A. J. Pomerans (New York, Harper and Row, 1971), 94-95.

[3] Robert Kargon, "Henry Rowland and the Physics Discipline in America," *Vistas in Astronomy*: special issue on *Henry Rowland and Astronomical Spectroscopy: Celebration of the 100th Anniversary of Henry Rowland's Introduction of the Concave Diffraction Grating*, ed. R. C. Henry, D. H. DeVorkin, and Peter Beer, 29, (1986), 131-136, on 131.

department finds "machinic" a good adjective for his physics.[4] Even Josiah Willard Gibbs, that most esoteric of theorists of thermodynamics and statistical mechanics, completed his Ph.D. at Yale in engineering, the first such doctorate awarded there. His dissertation was on nothing so distant from the everyday world as the atomic theory of matter but on a mathematical investigation of the optimal shape of gear teeth. This early training in engineering, however, has not yet been related to his well-known phenomenological and instrumental view of theory.

To get at the significance of engineering in American physics and of Rowland's place in its history, it is useful to consider the formal establishment of the discipline. On 20 May 1899 the American Physical Society was born at Columbia University in New York. The thirty-six men in attendance had responded to a letter from seven of their colleagues.[5] They promptly elected as their President and Vice President the two people who best represent the style of American physics in the process of its formation, Henry Rowland and Albert Michelson, neither of whom actually attended the meeting but who had agreed beforehand to lend their prestige to its offices. (Gibbs, never an organization man, supported the formation of the Society but declined to participate.) Like Rowland and Michelson—respectively, fourth generation heir to an establishment family of Presbyterian clergymen and immigrant Jew—the leading members were anything but uniform in their backgrounds. Michael Pupin, who hosted the meeting at Columbia, embodied the American success story, having transformed himself by hard work and brains from a lone sixteen-year-old Serbian immigrant in New York into an internationally recognized professor of mathematical physics. At the other end of the social spectrum, Arthur Gordon Webster, the real organizer of the meeting, often called the "father" of the Society, was the only son of a prosperous New England family with a Harvard College education.[6] And yet, across these cultural and social divides were many bridges.

One such bridge is evident in the fact that all of the organizers and many of the founding members had either taken their PhDs in Germany or studied there. Like Rowland, they typically studied in Berlin

---

[4] David C. Brock, "The Ryersonites: Chicago Physics and Institution-Building, 1875-1925" (dissertation, Princeton University, in preparation).

[5] Melba Phillips, "The American Physical Society: A Survey of its First 50 Years," *American Journal of Physics,* 58, no. 3 (1990), 219-230, reprints the letter and a list of those attending. Minutes of the early meetings are in *Bulletin of the American Physical Society* (published quarterly), ed. J. S. Ames, M. I. Pupin, Ernest Merritt, vols. 1-3 (1899-1902). The constitution, by-laws, and a complete list of members in 1902 is in vol. 3. Later minutes appeared in the *Physical Review,* when it was joined to the Society, beginning with 16 (1903), 173.

[6] Unless otherwise specified, biographical information is from the *Dictionary of Scientific Biography.* For Webster's important role, see Melba Phillips, "Arthur Gordon Webster, Founder of the APS," *Physics Today,* 40, no. 6 (1987), 48-52.

with Hermann Helmholtz (Rowland, Michelson, Pupin, Benjamin Osgood Peirce, Joseph Sweetnam Ames, Webster) but also in Berlin with Kirchhoff (Pupin) or Kundt (Webster) [Add M. F. Magie], in Heidelberg with Quincke (Michelson), in Leipzig with Wiedemann (Benjamin O. Peirce), or in Goettingen with [??] (Edward L. Nichols). Having gone to Germany in the first place to acquire the kind of specialized education in research that was still largely unavailable in the US, these men all returned with an increased ambition to professionalize their activities at home. Their regular complaints about poor laboratory facilities, inadequate mathematics education, and low intellectual standards should perhaps be read less as a measure of miserable conditions in the universities than of their united ambition for themselves and their discipline in a reorganized curriculum. A major move in that direction was the founding in 1893 of the *Physical Review: A Journal of Experimental and Theoretical Physics* by E. L Nichols and Ernest Merritt at Cornell, with the support of the university. Both Nichols and Merritt would soon play leading roles in founding the Physical Society.

The organizers also constituted a network in the private schools of the East Coast, several of them new but all with new ambitions for research in science: Hopkins, Princeton, Columbia, Cornell, Clark, Harvard, plus Chicago. And they were linked by personal friendship and by apprenticeship, especially to Rowland. Webster had served two years as Michelson's assistant at the new Clark University for graduate students in Worcester before taking over the physics department there when Michelson moved to head the new department at the University of Chicago in 1892. J. S. Ames made his entire career (undergraduate, graduate, colleague, and collaborator) with Rowland at Hopkins, the still-new flagship of research, while another collaborator, E. L. Nichols at Cornell, had done a post-doctoral year with him. Beyond the immediate organizers, Harvard's Trowbridge often acknowledged his debt to Rowland's department, acquired as a visiting professor there. His colleague Edwin S. Hall was one of Rowland's prize doctoral students. And at Yale, where an attempt to lure Gibbs away to Hopkins had failed, Rowland's assistant for eight years, Charles Hastings, carried the Hopkins message as department chair from 1884-1915.[7]

Thus the American Physical Society may be said to have been born as an "old-boy" network with the outspoken Webster as its organizing "father," Rowland its symbolic "father," and Michelson its symbolic "uncle." But it was a rapidly growing network which quickly expanded far beyond the old boys. Beginning with 38 mem-

---

[7] Kargon, "Rowland and the Physics Discipline," 135.

bers in May 1899, it had grown to 58 by December, to 144 in three years and to 282 in six years.[8] Such rapid growth speaks of high demand for physicists and requires once again that we see the role of Rowland and Michelson not so much as the creators of the discipline but as the inspiring models for a community whose rapid expansion and professionalization had sources reaching far beyond their own remarkable contributions. And this recognition leads immediately back to the engineering orientation of the new physics community as providing perhaps its most representative and unifying characteristic.

This view corresponds to the only extended account by a founding member that we have. Frederick Bedell was the younger colleague of Nichols and Merritt at Cornell, whom he joined as the third editor of the *Physical Review*. Bedell took the founding history of the society back to 1876, Rowland's first year at Hopkins and the year of the international exposition in Philadelphia celebrating the centennial of American independence. Above all he remembered Machinery Hall, dominated by a huge 1600 hsp Corliss engine but featuring also the first Edison light bulb, the first Bell telephone, and the first dynamo made in America, by William A. Anthony, professor of physics and electrical engineering at Cornell (later at Cooper Union as a founding member of the Society). These exhibits represented to Bedell a technological transformation of the culture. In retrospect, they also "typified the future of physics in industry and in education in which the Physical Society was destined to play its part." It was the excitement surrounding the inventions and the personalities of these men and others, such as Nikola Tesla and William Stanley, with their alternating current induction motors and transformers, that was driving the rapid growth and professionalization of physics. "The increasing amount of fundamental research at the universities, with increase of staff and equipment, was stimulating, and there arose a need for journals in specialized fields." And it was out of this demand for specialized technical knowledge, in Bedell's view, that the *Physical Review* was born at Cornell in 1893.[9]

In the same year the International Electrical Congress met in Chicago with delegates from all over Europe, including Helmholtz and a number of notable British and French physicist-engineers. Rowland served as President and Nichols as Secretary of the Chamber of Delegates while Webster chaired the Pure Theory Section. Bedell presented a paper and began lifelong friendships with Rowland and Webster. Thus their organizing activities and personal relations with respect to

---

[8] Frederick Bedell, "What Led to the Founding of the American Physical Society," *Physical Review*, 75 (1949): 1601-1604, on 1604.

[9] Ibid., 1601-2.

electrical engineering presaged those for the Physical Society.[10] Of the mathematically inclined Webster, J. S. Ames observed, "He was as much interested in what one may properly call the engineering side of his subject as in the purely physical one, and his ability was so great that there was no practical field in which he could not venture with great profit to all concerned."[11] Similar associations can be drawn for others among the founding members. Notable is Pupin at Columbia, who returned from Berlin in 1889 to teach mathematical physics in the newly formed department of electrical engineering, where he became professor of electromechanics in 1901. He achieved fame for his mathematical and material inventions in telegraphy and telephony and registered 34 patents during his lifetime.

To recognize the prominence of electrical and mechanical engineering in the physics of the late 19th century is to obtain a somewhat different view than has been usual in the historical literature. We have available two extended treatments; Daniel Kevles's *The Physicists* (1978), still by far the most useful and interesting survey available, and Albert Moyer's *American Physics in Transition* (1983), with a more specialized philosophical aim. For different reasons neither Kevles nor Moyer focused attention on the centrality of machines in the everyday lives and intellectual commitments of American physicists, though both were fully aware that the growth of physics in the universities depended on its service to engineering. Kevles was primarily interested in the institutional development of the new profession by comparison with its established European counterpart. Not unnaturally he contrasted the largely experimental work of the Americans with the theoretical work of many of their brethren overseas. Under this contrast American physicists lacked the capacity for research in mathematical and theoretical physics and therefore limited themselves to experimental and practical work, pursuing a kind of Baconian fact-gathering while reacting to European ideas.[12] This characterization, though correct in a sense, misses the point that the people who did pursue mathematical and theoretical physics did so with an engineering bent and with great confidence in the worth of their enterprise. Webster and Pupin are representative; even the anomalous Gibbs may qualify. Robert Kargon remarks that "American physics seems to have cultivated a curious type of borderland physics-engineer hybrid, not unlike Rowland himself."[13] Or as Hoag put it in 1929 to Heisenberg,

---

[10] Ibid., 1603.

[11] Phillips, "Webster," 48.

[12] Daniel J. Kevles, *The Physicists: The History of a Scientific Community in Modern America* (New York, Knopf, 1978), e.g., chapter V, "Research and Reform."

[13] Kargon, "Rowland and the Physics Discipline," 136.

even American theoretical physicists "behave just like the engineer building a new bridge."

Heisenberg's discussion of Hoag's remark goes off in a rather philosophical direction, which is another way to discuss interesting subjects while missing the significance of American physics riding on engineering. Albert Moyer has written a perceptive survey of the changing philosophical positions of late nineteenth century American physicists, judging their commitments according to their degree of adherence to or distance from what J. B. Stallo called "atomo-mechanism" in his 1882 critique of modern physics. In this mode of analysis, their empiricist and antimetaphysical sentiments, which with few exceptions were quite unsophisticated philosophically, appear as forms of phenomenalism, operationalism, or positivism rather than attitudes grounded in everyday practice. Like the experiment-theory axis, this one yields summary statements of the fact-gathering sort: "And most [pre-1900 physicists]—Michelson in his ether studies and Rowland in his measurements of the mechanical equivalent of heat—were dedicated to routinely expanding the data base of received mechanical theories."[14] Again, there may be some justice in treating American physics as almost entirely experimental, particularly when seen as an answer to the question of why American physicists contributed little to the new theories of relativity and quantum mechanics yet readily accepted them.

But "expanding the data base" does not capture the self-confident sense of capacity, even muscularity, that animated so many Americans during their great technological and industrial expansion. Rowland never seems to have suffered a moment's doubt about his own abilities or his enterprise. When he traveled in Germany in 1876 examining laboratory arrangements and instruments for the projected Hopkins laboratory, his comments were as often caustic as admiring. He and his peers brought back from Germany much of value for physics research, but their imports were highly selective. They discarded anything they considered metaphysical speculation, such as atomic theory. Certainly they were playing catch-up, but they were playing by their own rules. Such independence reinforces the view that to a considerable degree they shared a generic set of values rooted in their common appreciation of instrument-based physics, their personal skill in making mechanical and electrical devices, and their commitment to precision measurement as the vehicle of progress. From this more positive perspective, it might well be said that the diffraction gratings and solar wavelength standards that Rowland distributed throughout

---

[14] Albert E. Moyer, *American Physics in Transition: A History of Conceptual Change in the Late Nineteenth Century* (Los Angeles, Tomash, 1983), 118.

Europe (while keeping his ruling engine secret) were as important to the development of quantum mechanics as, say, J. J. Thomson's theory of the atom.

George Sweetnam does not write about the issues at this broad untextured level, which he would have regarded as too distant from the sources. Instead, he investigates Rowland's actual work in what he calls "laboratory engineering." It was characterized in the first instance by an incomparable ability to invent simple experiments of great sensitivity to reveal physical effects which one could imagine theoretically to exist but which seemed to defy observation. These experiments began with his early attempts in the 1870s to analyze the magnetization of materials in response to a magnetizing force, which after adroit mathematical simplification yielded an "Ohm's law" of magnetization (albeit without hysteresis).[15] With support from Maxwell they also yielded publication in the widely read *Philosophical Magazine* and the offer in 1875 from Daniel Coit Gilman, President of Hopkins, to head the new research laboratory there. Rowland's electromagnetic experiments continued with his remarkable demonstration, carried out while working in Helmholtz's laboratory, of electromagnetism produced by convection of static charge. That accomplishment immediately established his international reputation. It was followed in 1879 by the equally famous demonstration, carried out by his student E. H. Hall at Hopkins, of the effect of a magnetic field on a current flowing perpendicular to it, the "Hall Effect."[16]

These experiments of sensitivity were matched by Rowland's experiments of precision. His first major project at Hopkin's after returning from his European tour was a redetermination of the mechanical equivalent of heat. He had learned to regard this number as the single most important constant in the new energy physics that had been developing since mid-century and one of the key elements in the attempt to establish an international system of units based on the exchange relations connecting the various branches of physics: mechanics, heat, electricity, magnetism, and light.[17] Ultimately even more important to Rowland in this enterprise of quantitative unification was measuring the wavelengths of the unique spectral lines emitted by "atoms," what-

---

[15] John D. Miller, "Rowland's Physics," *Physics Today*, 29, no. 7 (1976): 39-45.

[16] Jed Z. Buchwald, *From Maxwell to Microphysics; Aspects of Electromagnetic Theory in the Last Quarter of the Nineteenth Century* (Chicago, University of Chicago Press, 1985): 73-95, gives extended discussion of the discovery of the Hall Effect.

[17] H. Otto Sibum, "An Old Hand in a New System," in Jean-Paul Gaudillere, Illana Lowy (eds) *The Invisible Industrialist: Manufactures and the Production of Scientific Knowledge* (London, Macmillan, 1998), pp. 23-57. Sibum is completing a book on measurements of the mechanical equivalent of heat which will include an extended discussion of Rowland's work, its sources and context.

ever they might be, and showing that these markers were the same throughout the terrestrial and celestial realms. Using prism spectroscopes to separate the spectral lines, Bunsen, Kirchhoff, and others had established this project in the 50s and 60s. A. J. Ångström used plane diffraction gratings in the 1860s to obtain some precision in measuring wavelengths and at least two American acquaintances of Rowland make progress in perfecting the technique.[18] But it was Rowland himself who astonished the world with his concave diffraction gratings and standardizing solar spectra. It is of this work and the school of physics which grew up around it and beyond it that Sweetnam writes. Among his most significant findings is the role played by the diffraction gratings themselves, in all their material and mechanical concreteness, as carriers of the school.

Sweetnam's focus on the material means and the personal motivation for attaining sensitivity and precision draws out two features of American physics to which recent scholarship has paid increasing attention: cultural values and technology. On the values side, laboratory discipline, constant pursuit of sources of error, and avoidance of speculation were widely represented not only as imperatives for research but as exemplary moral imperatives, as models for building the character traits of humility, integrity, and strength, or what one used to call the protestant ethic. Rowland, descended from three successive generations of Presbyterian ministers, advertised this idealistic vision of physics in numerous lectures, most famously in his "Plea for Pure Science" delivered in 1883 to the AAAS and continuing through his Presidential address at the second meeting of the American Physical Society in 1899. But the moral ideal as he and other scientists advocated it was grounded in practical laboratory work with materials and machines, in engagement with the real world. In a nearly seamless manner it merged the conditions for moral leadership in society with those for material progress. "Nearly" is an important qualifier, for it became crucial to separate crass material motives, or profit, from the pure quest for knowledge of matter, which might well yield profits when exploited. This tenuous merger of morality and matter required considerable readjustment in the values of elite universities like Harvard, Princeton, and Yale in order to incorporate into their traditional religious and intellectual mission the science and engineering school that they had so consciously kept separate. The problem was by no means unique to America. Maxwell faced much the

---

[18] Deborah Jean Warner, "Rowland's Gratings: Contemporary Technology," *Vistas in Astronomy*: special issue on *Henry Rowland and Astronomical Spectroscopy; Celebration of the 100th Anniversary of Henry Rowland's Introduction of the Concave Diffraction Grating*, ed. R. D. Henry, D. H. DeVorkin, and Pter Beer, 29 (1986), 125-130.

same challenge at Cambridge University when in 1873 he began to integrate the workshop-like atmosphere of the new Cavendish Laboratory into the gentlemanly traditions of the university.[19] In Germany great school debates raged through the end of the century over modernizing the classical *Gymnasia*. But the merger seems to have found more fruitful ground in the US through the technological and managerial accomplishments that were powering both cultural optimism and industrial expansion.

As Sweetnam and others show, the sensitivity and precision of Rowland's "laboratory engineering" depended not only on his own considerable talent as an inventor but on his intimate familiarity with the practical methods of contemporary machine building and on his access to people even more involved than he.[20] His ruling engine for diffraction gratings would have been inconceivable without the very recent developments in the machine tool industry that had increased the accuracy and precision of machining by two orders of magnitude and that, along with rigorous oversight and quality control, had made possible the mass manufacture of interchangeable parts in the "American system."[21] Indeed, his own ruling engine built directly on that of William A. Rogers, the Harvard astronomer best known for his work with George M. Bond of Pratt and Whitney on the Rogers-Bond comparator, which set the standard in the machine tool industry with an accuracy of one millionth of an inch.[22] In addition he relied heavily on his master mechanic Theodore Schneider for the construction, maintenance, and operation of the engine and on the precision instrument maker John A. Brashear to supply the speculum metal blanks on which the gratings were ruled. In short, the professor, his mechanic, his machines, and his materials all expressed in their various ways the technological-industrial revolution taking place in America.

From this perspective, and taking Rowland as representative, the practical emphasis of American physicists, their rejection of theoretical speculation in favor of a pragmatic use of theory, concrete mechanical

---

[19] Simon Schaffer, "A Manufactory of Ohms: Late Victorian Metrology and its Instrumentation," in Susan Cozzens and Robert Bud, eds. *Invisible Connexions* (Bellingham: SPIE, 1992), 23-56. Idem., "Accurate Measurement is an English Science," in M. N. Wise, ed., *The Values of Precision* (Princeton, Princeton University Press, 1995), 135-172.

[20] George Sweetnam, "Henry Rowland, the Concave Diffraction Grating, and the Analysis of Light," in M. N. Wise, ed., *The Values of Precision* (Princeton, Princeton University Press, 1995), 283-310.

[21] Paul Uselding, "Measuring Techniques and manufacturing Practice," in Otto Mayr and R. D. Post, eds, *Yankee Enterprise: The Rise of the American System of Manufactures* (Washington, D.C., 1981), 103-126; and in the same volume, David A. Hounshell, "The System: Theory and Practice," 127-152.

[22] Warner, "Rowland's Gratings" shows that all ruling engines in the 19th century were adapted directly from dividing engines used in making precision machinery and scales.

models, and precision measurement, looks like a preference for exploiting resources in which they possessed an evident advantage while devaluing theoretical and philosophical traditions which they lacked. Certainly the decision to pursue precision measurement as the most productive route into the future of physics was a conscious one for both Rowland and Michelson and they organized their laboratories around it. Their practicality was not of the chewing gum and sealing wax variety but of cutting edge technologies and professionally crafted instruments.

Again from this perspective, Rowland's repeated pleas for pure science look as much like a fear that his colleagues could all too easily turn their enterprise into an Edisonian science of profits as it was a critique of Edison's own activities.[23] That is, the pressing need to demarcate pure from applied science seems to have arisen simultaneously with the recognition that the requirements for technological development and for research science were much the same and with the threat that the research laboratory might well be mistaken for a factory, both architecturally and intellectually.

George Sweetnam's study of *The Command of Light* explores the everyday practices and attitudes that characterized the Hopkins school of physics. It thereby raises issues of quite general import about Rowland's representative position in American physics. These issues will provide ground and stimulation for many further such studies.

---

[23] David A. Hounshell, "Edison and the Pure Science Ideal in 19th Century America," *Science*, 207 (1980): 612-617.

Engineering a New Discipline

The house of Rowland may long have dreamed of reconciling Heaven and Earth, mediating as they did between the sublime and the mundane, between Powers above and powers on terra firma. The nineteenth century saw three Henry Augustus Rowlands, in lineal descent: two ministers, and one physicist. The physics profession was new in the time of Henry the third, as new as the research university in America. Yet the position of the physicist, like that of the first two Henrys, required him to be an exemplar to those around him; in his words, a physics educator had to be exemplary so that his students could "see before them this high and noble life,"[1] in which the dispassionate search for truth was paramount. And the heavens, although not necessarily Heaven, became part of the physicist's domain, too.[2]

The human condition, he wrote, required each individual "to discriminate between right and wrong, between truth and falsehood." The exceptional man, especially aware of the possibilities of error, had the courage nevertheless to maintain his opinions: "Like Galileo and Copernicus, he inaugurates a new era in science, or like Luther, in the

---

[1]Henry Augustus Rowland [III], "A Plea for Pure Science" (1883), in *The Physical Papers of Henry Augustus Rowland (PPR)* (Baltimore: The Johns Hopkins University Press, 1902), 598.

[2]On reconciling Heaven and Earth, Henry Augustus Rowland [I] stated: "The devout exercise of praise, in earthly communities, will bring them to a near resemblance to the blessed society above." *A Discourse, Delivered November 27th, 1800, A Day Observed as an Anniversary Thanksgiving* (Hartford: Hudson and Goodwin, 1801), 6.

religious belief of mankind."[3] Like these revolutionaries, the nineteenth-century physicist pursued truth by a process of testing and inquiry. The professional methodology did not wholly differ from that of Henry the second, who believed that a minister "should know experimentally what true religion is."[4]

There was no quarrel between religion and science. Henry the first believed that America was specially favored: "Here science has diffused her cheering beams."[5] Henry the second even held a strong amateur interest in science, which carried over to his son,[6] and he adumbrated a physicist's work when he maintained that the Deity "gave his disciples no power over persons, but things."[7] Henry the physicist (1848-1901) excelled in the control of things, and therein lay his principal claim to eminence in the American profession which he helped to create, before the turn of the twentieth century.[8]

From the time of his thesis on the steam engine at Rensselaer Polytechnic Institute in 1870, until his invention of a revolutionary spectroscope in 1882, there were but few indications that light would become a principal subject of Rowland's control. In 1870, he still emphasized fundamental principles:

> The central idea of modern science is force. Of this force there is supposed to be a certain quantity in the universe which can be neither increased or [sic] diminished.[9]

The doctrine of the conservation of force or energy was fairly new in science, having existed for only twenty to thirty years,[10] so the essay demonstrated the student's awareness of emerging concepts. As a student of civil engineering in upstate New York in the late 1860s, Row-

---

[3]Rowland [III], "The Physical Laboratory in Modern Education" (1886), in *PPR*, 614.

[4]Henry Augustus Rowland [II], *The Excellency of Our Christian Polity* (New York: M.W. Dodd, 1851), 5.

[5]Rowland [I], *A Discourse*, 16.

[6]A biographer of the physicist stated, "From his father he inherited his love for scientific study." Thomas C. Mendenhall, "Henry Augustus Rowland, 1848-1901," *Biographical Memoirs of the National Academy of Sciences (BMNAS)*, 5 (1905): 127.

[7]Rowland [II], *The Excellency*, 26.

[8]In the physicist's day, the Swiss scientist Alphonse de Candolle remarked that European Protestants were more likely than Catholics to become scientists. But the effect was even more specific: "Remove from the lists of scientists of Protestant countries the sons of ministers, and equality would almost be reestablished between the populations of the two sects from the point of view of their influence on science." He conjectured that the celibacy of priests was one reason for the actual difference. De Candolle, *Histoire des Sciences et des Savants depuis Deux Siècles* (Geneva: H. Georg, 1885), 330, 334.

[9]Rowland [III], "Steam Engine with Variable Cutoff" [ms.], Series 5, Box 41, Henry Augustus Rowland Papers (HAR), Ms. 6, Special Collections, Milton S. Eisenhower Library, Johns Hopkins University Press, Baltimore, MD.

[10]The need for imprecision in attributing a date to the principle is clear from Thomas Kuhn, "Energy Conservation as an Example of Simultaneous Discovery," in *The Essential Tension* (Chicago: University of Chicago Press, 1977), 66-104.

land carried out many experiments independently of his coursework. He assembled historical data obtained by others on the aurora borealis (the northern lights), sunspots, and terrestrial magnetism, evidently hoping to find some connection,[11] but he undertook no major investigations of sunlight. Later, in 1873, he studied the spectrum of the aurora, but his published report was very brief.[12]

Rowland's education and independent research in physics began on his own initiative, continued while he studied civil engineering, and still progressed when he began teaching physics to future engineers at Rensselaer. His approach centered on apparatus that he could build himself. Progress in physics meant, to him, coming to grips with nature in the laboratory, often by mechanical means. Thus training in engineering was not extraneous; it became an integral part of his style of doing physics.

At Rensselaer, Rowland's intellect roamed well beyond the formal curriculum. A colleague later described his explorations:

> He made immediate use of the opportunities afforded in Troy and its neighborhood for the examination of machinery and manufacturing processes, and one of his earliest letters to his friends contained a clear and detailed description of the operation of making railroad iron, the rolls, shears, saws, and other special machines being represented in uncommonly well executed pen drawings.[13]

Machines were transparent to Rowland; he could trace their inner workings in his mind. They were part of the landscape. Later, in the pursuit of physics, Rowland still thought in a concrete language of machines and other devices.

When the instruments at Rensselaer were not adequate for his experiments, he personally compensated, bringing some instruments from his former home in Newark, New Jersey; making others with the few tools available. Still others he built and maintained in his room in a boarding house. At times the only spaces for experiment were the bed and washstand in his own room. As a colleague later observed, "He spent all of his leisure in designing and constructing physical apparatus of various kinds with which he experimented continually."[14] Soon he reported back to Newark that he had built a balance, a galvanometer, an electrometer, an induction coil, and a Ruhmkorff coil. Rowland also constructed a small steam engine which would power other experiments. He observed the colored light given off when electricity was discharged through tubes of gas. In 1867, he began to keep

---

[11]"Volume 1," 1868 notebook, 112-113, Box 20, HAR.
[12]"On the Auroral Spectrum" (1873), in *PPR*, 31.
[13]T.C. Mendenhall, "Henry Augustus Rowland," 129.
[14]Ibid., 130.

records of his experiments in Troy.[15] He also constructed a "Galvanic battery," a polariscope, a furnace for melting iron, a "Delicate ballance [sic]," vacuum tubes, and other devices.[16]

While still an engineering student, Rowland decided to devote himself to science.[17] At the time, he had published only one scientific paper, a short letter to the editors of *Scientific American*, on the formation of vortices in water draining through an outlet.[18] His decision to pursue physics was not obvious or easy; he carried out much of his experimentation on his own. After graduating from Rensselaer, he performed experiments for a while in his home, before finding work as a railroad surveyor.[19]

A distinctive mark of Rowland's physics, apparent already in his undergraduate years, was the need he felt for instruments, and his willingness to build them himself if necessary. After his employment as a railroad surveyor, he became an instructor in physics at the College of Wooster in Ohio, but experienced tension with administrators, who felt that he spent too much money on scientific equipment. He subsequently returned to Rensselaer as an instructor and then assistant professor in physics. He obtained an agreement from the Institute to devote hundreds of dollars to the purchase of laboratory apparatus, or, failing that, to add to his pay. When the funds for apparatus never arrived, Rowland insisted on the higher pay. But even as a faculty member, he found that the Institute provided no adequate space, nor much encouragement, for his laboratory research. In the winter of 1874-1875, he worked in a small shed, where his breath froze on the instruments. Later, he obtained the use of a room with an iron stove, but the iron in the stove affected magnetic measurements. Rowland then opened negotiations for a possible position at the University of Pennsylvania.[20] (Despite Rowland's difficulties, the official posture of Rensselaer was favorable, not hostile, to science. An 1838 alumnus said of the larger facilities in Rowland's time: "I have never seen [a laboratory] more complete in all its arrangements and appliances.")[21]

---

[15]Ibid., 130. John David Miller, "Henry Augustus Rowland and His Electromagnetic Researches" (Ph.D. dissertation, Oregon State University, 1970), 12-13, 15, 354.

[16]Rowland, "Note or Receipt Book," 38, Box 20, HAR.

[17]T.C. Mendenhall, "Henry Augustus Rowland," 131.

[18]"The Vortex Problem" (1865), in *PPR*, 23.

[19]T.C. Mendenhall, "Henry Augustus Rowland," 118.

[20]Miller, 80-81, 87, 355. T.C. Mendenhall, 131.

[21]Norman Stratton, "Address," in *Proceedings of the Semi-Centennial Celebration of the Rensselaer Polytechnic Institute* (Troy, NY: Wm. H. Young, 1875), 90. Another alumnus went further, and suggested that methods of instruction at Rensselaer had long been parallel to methods at such European institutions as the laboratory of German chemist Justus von Liebig. E.N. Horsford, "Address," ibid., 49-50. Robert Bruce reports, "As a Rensselaer graduate Horsford had known a teaching lab before, but Liebig's reached a new level and scale." *The Launching of Modern American Science* (Ithaca: Cornell University Press, 1988), 23.

In 1875, while Rowland was struggling to establish a secure place for physics at Rensselaer, The Johns Hopkins University in Baltimore began to emerge from an idea and an endowment into an organization. As it turned out, this provided just the opportunity that Rowland needed. By the time Daniel Coit Gilman, president of the new university, began his search for faculty, Rowland had published about ten articles, including short letters. One paper described a mechanical model, using coupled pendulums, that demonstrated resonance phenomena.[22] Another paper, mentioned above, suggested the presence of calcium in the aurora borealis, based on the auroral spectrum.[23]

James Clerk Maxwell, the Scottish theoretician of electricity and magnetism, had recognized the value of Rowland's more mathematical work, and arranged for publication of some in the British *Philosophical Magazine*—a breakthrough in Rowland's career. One view holds that "there is evidence to suggest" that American editors were "not comfortable with dissertations involving exact mathematical methods, even when it could get them." In this view, the American editors could not understand Rowland's mathematics, whereas Maxwell could. This analysis would agree with a broader assertion that American scientists prior to the 1930s were deficient in mathematics, relative to their European counterparts. Thus Rowland may have faced incomprehension in his own country.[24]

His situation may have resembled, to a small degree, that of Yale physicist Josiah Willard Gibbs (1839-1903), the most brilliant exception to the historical generalization that nineteenth-century American scientists were indifferent to theory. Gibbs, who received the Ph.D. from Yale in 1863 for a dissertation on the optimal shape of gear teeth, cultivated vector analysis, chemical energetics, physical chemistry, and statistical mechanics. Yet he was little understood among his compatriot contemporaries. A contemporary biographer concluded that, for Americans, "really important work may be better appreciated abroad than at home." Gibbs received no salary from Yale until after he received an offer from Johns Hopkins, at Rowland's prompting, in 1879. The Hopkins physicist is said to have called the counter-offer from Yale "the greatest crime of the century," because it kept Gibbs away from Johns Hopkins.[25]

---

[22]"Illustration of Resonances and Actions of a Similar Nature" (1872), in *PPR*, 28-30.

[23]"On the Auroral Spectrum," 31.

[24]Miller, dissertation, 53-56. On the question of mathematical proficiency among U.S. scientists, see John Servos, "Mathematics and the Physical Sciences in America, 1880-1930," *Isis*, 77 (1986): 611-629.

[25]Edwin E. Slosson, "Willard Gibbs, Physicist, 1839-1903," in *Leading American Men of Science*, ed. David Starr Jordan (New York: Henry Holt and Co., 1910), 348, 353.

Compared to Gibbs, Rowland was not a theoretician, and he never demonstrated the mathematical creativity that Gibbs displayed in developing vector analysis and sophisticated techniques for statistical mechanics. Rowland's use of mathematics was relatively moderate. In a formal sense, the most daunting aspects of his equations, apart from the inclusion of many terms, were the use of integrals, the use of the "Napierian logarithmic base" (base e), and the use of power series.[26] However, the ideas expressed by the equations may have been unfamiliar as well, and may provide an alternative explanation for the *Journal* rejections.

In a paper published eventually in Britain, Rowland claimed that his experiments were the first "in which the results are expressed and the reasoning carried out in the language of Faraday's lines of magnetic force." The appeal of Faraday's language was not in formal nicety. Rather, the advantages were practical: it simply worked. As Rowland said, "Whether Faraday's theory is correct or not, it is well known that its use will give correct results; at the present time the tendency of the most advanced thought is *toward* the theory." From William Thomson, Rowland drew on analogies that would clarify his equations, for "the mathematical treatment of magnetism is the same as that of the flow of heat in a solid, as the static induction of electricity, and as the flow of a frictionless incompressible liquid through a porous solid." And from Maxwell, he drew one further analogy, with the conduction of electricity.[27] These ideas together may have been alien to American reviewers, but Rowland's understanding of electricity and magnetism was not confined by national borders, and the level of his work thus required, from early on, his involvement in an international community of physics.

The most important thing the American gained from Maxwell was recognition, which he could not fully acquire in his own country. Thus internationalism, in the sense of a physics discipline unconfined by national boundaries, was another mark of his style of physics. From Britain, his reputation made its way back to America, where he might have waited long indeed, without it. After he published twice on magnetism in the *Philosophical Magazine*, the editors of the *American Journal of Science* finally accepted an article on magnetism for publication.[28]

---

[26]These appear in the paper eventually published as "On Magnetic Permeability, and the Maximum of Magnetism of Iron, Steel, and Nickel" (1873), in *PPR*, 35-55.

[27]Ibid., 37-38, 40. Emphasis in original. On Thomson's use of analogy, see Crosbie Smith and M. Norton Wise, *Energy & Empire: A Biographical Study of Lord Kelvin* (Cambridge: Cambridge University Press, 1989), 263-275.

[28]The second *Philosophical Magazine* article was: "On the Magnetic Permeability and Maximum of Magnetism of Nickel and Cobalt" (1874), in *PPR*, 56-74. The paper on magnetism for the *American Journal of Science* was "On a New Diamagnetic Attachment to the Lantern, with a Note on the Theory of the Oscillations of Inductively Magnetized Bodies" (1875), in *PPR*, 75-79.

Maxwell's recognition held the implication that, when Daniel Coit Gilman began his search for faculty for Johns Hopkins, Rowland had a name, and Gilman heard of his work in 1875 while visiting the U.S. Military Academy at West Point. Rowland was called there, conversed with Gilman, and soon had an appointment as professor of physics at the new university, to open in one year—giving him time to travel in Europe at the expense of Johns Hopkins and become acquainted with the most advanced work in physics.[29] Rowland was suddenly free of the skepticism with which Rensselaer administrators had regarded his experiments, and he was empowered to purchase the latest apparatus and instruments in Europe, as well. Gilman fully expected him to devote considerable time to experimental research.

During his European travels, Rowland compiled a list of ninety-six pieces of apparatus suitable for his planned physics department. Even though the total cost amounted to $6,429, he received authorization from his generously endowed university to buy it all. (Rowland's annual salary as of April 1876, by comparison, was $3,000.)[30] His new employer was glad to subsidize the equipment that two previous institutions had greatly restricted. As a result, Johns Hopkins soon had, in at least one appraisal, the best-equipped physics laboratory in the U.S., and one of the best-equipped in the world.[31] Doing good physics was partly a matter of having the best tools. But there could be tools that money could not buy, and in some cases it remained for Rowland or his students to devise them.

Probably the crowning experience of the European trip was a chance to work in the laboratory of the great German master, Hermann von Helmholtz, where numerous young physicists found inspiration. Rowland's most celebrated single experiment, conducted in Helmholtz's Berlin laboratory in 1876, measured the magnetic effects of a rotating, electrified disk. He had conceived the experiment as a way of determining "whether or not an electrified body in motion produces magnetic effects." Characteristically, he addressed the issue by engineering an apparatus that might resolve the issue directly. He wrote, "There seems to be no theoretical ground upon which we can settle the question." After operating the apparatus, he concluded that the experiment confirmed Maxwell's idea that electric charge in motion, a "convection current," produces a magnetic field.[32] The conclusion was

---

[29]T.C. Mendenhall, "Henry Augustus Rowland," 132. Daniel Coit Gilman, *The Launching of a University* (New York: Dodd, Mead & Co., 1906), 14-15.

[30]Miller, dissertation, 185, 188, 190.

[31]Daniel J. Kevles, "Rowland, Henry Augustus," Vol. 11 (1975), *Dictionary of Scientific Biography* (*DSB*), ed. Charles Coulston Gillispie, 578.

[32]"On the Magnetic Effect of Electric Convection" (1878), in *PPR*, 128-129. This article originally appeared in the *American Journal of Science*.

widely discussed, and Maxwell celebrated the experiment in verse the following year:

> *The mounted disk of ebonite*
> *Has whirled before nor whirled in vain,*
> *Rowland of Troy, that doughty knight,*
> *Convection currents did obtain,*
> *In such a disk, of power to wheedle*
> *From its loved north the subtle needle.*[33]

The experiment received significant attention, even though doubts about the results arose in some quarters, where replication failed to corroborate Rowland's findings.[34] He had formulated the experiment in correspondence with Helmholtz, in which he asked for space in Berlin laboratory to carry it out. He believed, however, that after the experiment was completed, physicists often attributed his own results to Helmholtz. Although the German physicist had suggested modifications in the apparatus, Rowland subsequently claimed for himself "*all* the credit for ideas, design of apparatus, the carrying out of the experiment, the calculation of results, and *everything* which gives the experiment its value." The original idea for the experiment, he noted, occurred to him in 1868, long before he met Helmholtz.[35] (In 1875, the German physicist had performed a somewhat different experiment that involved rotating the plates of a cylindrical capacitor in a magnetic field, which caused an electromotive force in the plates. Like Rowland's experiment, it produced results that tended to favor Maxwell's ideas.)[36]

But the provision of materials, and even more importantly the provision of space in a recognized center of physics, did help him. By moving in the international community, Rowland likely received far more notice for his work than if his experimental context were limited to Troy, New York. Like Maxwell, Helmholtz introduced the American to a larger community, more capable of appreciating his work. As a nineteenth-century American, Rowland could not conduct notable research without construing his domain as international.

---

[33]Quoted in "Prof. H.A. Rowland" [obituary], *Nature*, 64 (1901): 16.

[34]T.C. Mendenhall, "Henry Augustus Rowland," 120-121. Henry Crew, a student of Rowland in the 1880s, later recounted the doubts that arose: "Certain European experimenters failed to reproduce the phenomenon. Then in 1888 Rowland + Hutchinson repeated the expt. in Baltimore. Then, about 1900 Cremieu failed to find it in Paris; accordingly Harold Pender went over to Paris from Rowland's laboratory and showed Cremieu just where his mistake was." Henry Crew to Woodbury, 4 December 1935, Henry Crew Papers (HC), Northwestern University Archives, Evanston, IL.

[35]"Note on the Magnetic Effect of Electric Convection" (1879), in *PPR*, 138. Emphasis in original.

[36]R. Steven Turner, "Helmholtz, Hermann von," *DSB*, Vol. 6 (1972), 252.

During visits to European departments of physics, he paid particular attention to arrangements for instruction, especially laboratories for teaching, and lecture halls. He often noticed, and even sketched, the arrangements for indoor illumination. He visited the area at the Royal Institution in London where Michael Faraday had worked, and found "a small dark room needing gas in the daytime." Another room had a skylight that could be blocked by moving a horizontal screen. A lecture hall in Munich simply used window curtains to regulate light. There, Rowland saw a physicist analyze light while giving instruction: "He throws the diffraction spectra on a screen and afterwards measures the distances, and so gets the wave length in a lecture."[37] This demonstration may have impressed itself upon Rowland's memory, for he never tired in later years of proclaiming the ease of using diffraction instruments to measure wavelengths, compared with the difficulties of using prisms.

He also took interest in evaluating the proper combination of teaching and research. At University College in London he encountered the principle that "all advanced teaching in physics should be done by work in the laboratory with occasional lectures on *special points.*" The result was greater breadth of knowledge; experience had shown that students facing examinations, rather than laboratory work, limited the scope of their studies. At University College, one physicist told Rowland "that the system of the examinations makes it very hard for him to get students to study anything that was not *set* and he has no doubt that his highest men on examination are those who have confined themselves to the narrow gauge prescribed in the examination."[38]

Original research required the development or acquisition of innovative apparatus, which was an integral part, a *sine qua non*, of doing physics. After his move to Johns Hopkins, Rowland's noteworthy studies included the absolute value of the ohm;[39] the supervision of his student Edwin Hall in exploring the phenomenon which became known as the Hall effect;[40] the development of diverse ways of measuring electrical quantities;[41] and a major study of the mechanical

---

[37]Notebook - "European tour (1875)," 8, 15, 77, Box 22, HAR.

[38]Ibid., 12-13. Emphasis in original.

[39]"Research on the Absolute Unit of Electrical Resistance" (1878), in *PPR*, 145-178. "The Determination of the Ohm" (1884), ibid., 217-218. "The Value of the Ohm" (1887), ibid., 239-240.

[40]See Jed Z. Buchwald, *From Maxwell to Microphysics* (Chicago: University of Chicago Press, 1985), 73-95.

[41]"Electrical Measurement by Alternating Currents" (1897), in *PPR*, 294-313. Rowland and Thomas Dobbin Penniman, "Electrical Measurements" (1899), ibid., 314-337.

[42]"On the Mechanical Equivalent of Heat, with Subsidiary Researches on the Variation of the Mercurial from the Air-Thermometer and on the Variation of the Specific Heat of Water" (1880), in *PPR*, 343-468.

equivalent of heat.[42] His physics always revolved around concrete, specific laboratory apparatus, and thus always entailed laboratory engineering. As in his self-education and his initiation into Helmholtz's laboratory, the Baltimore physicist did not enter an investigation until he had seen how to engineer an apparatus that would directly address a significant question.

In the first six years of the Johns Hopkins physics department, a few lines of inquiry touched on light. From data in the rotating-disk experiment, Rowland derived "the ratio of the electromagnetic to the electrostatic system of units": 300,000,000 meters per second, the velocity of light.[43] This type of computation was not new; only the underlying measurements were. Rowland also returned to the subject of the northern lights, suggesting electrical charge circulating in the atmosphere as a causative agent: "[W]ere the earth electrified, the electricity would be carried to the higher latitudes by convection, would there discharge to the earth as an aurora, and passing back to the equator would get to the upper regions as a lightning discharge, once more to go on its unending cycle." This electric charge "would tend to circulate in the same way as the air from the equator to the poles and conversely." In Paris, he proposed arranging "on the whole earth, and especially in the polar regions, a systematic series of observations on atmospheric electricity."[44]

Besides the northern lights and new physical understandings of the velocity of light, there was a new technology attracting considerable public attention: Thomas Edison's electric light bulb of 1879. The device was not, by several decades, the first to produce light from electricity, but it was the first to hold promise for widespread domestic use, and Rowland set out to assess that potential. Together with George Barker of the University of Pennsylvania, Rowland tested the efficiency of Edison's invention, to measure the amount of light produced per horsepower of energy consumed. The study commenced, they said, "because most of the information on the subject has not been given to the public in a trustworthy form." Thus the two stepped forward to represent disinterested science for the public. Edison, they noted, placed "his entire establishment at our disposal." Their investigation found that the lamps could produce 100 to 200 "candles" per unit of horsepower, and even greater efficiencies might be achieved. The study concluded that Edison's light would indeed be practical,

[43]"On the Magnetic Effect," 137.

[44]"On Professors Ayrton and Perry's New Theory of the Earth's Magnetism, with a Note on a New Theory of the Aurora" (1879), in *PPR*, 183. Also, "On Atmospheric Electricity" (1881), ibid., 214, 215. In the first paper, Rowland also stated his belief that "the magnetism of the earth still remains, as before, one of the great mysteries of the universe, toward the solution of which we have not yet made the most distant approach," 182.

after "further experiment," provided that it could be made "cheap enough or durable enough."[45]

The northern lights, the electromagnetic derivation of the velocity of light, and the light bulb constituted the sum total of Rowland's published interest in light prior to 1882. The process of invention that followed, and that revolutionized the analysis of light, remains somewhat enigmatic, deeply rooted as it was in the engineering cognition of one physicist; he had pondered diffraction instruments privately as early as 1876.

Ira Remsen, Rowland's friend and colleague on the Johns Hopkins faculty, was apparently present, six years later, for the moment of epiphany. He recollected that, as the two rode by train from Baltimore to Washington, for a meeting of the National Academy of Sciences, an idea coalesced: "Just before we reached Washington he threw up his hands and said, 'It will work. I'm sure of it.' These words were not addressed to me but to space." Rowland promptly dropped the meeting from his schedule and caught the next train back to Baltimore, to work out his ideas.[46] What emerged was a revolutionary means of analyzing light—not producing it, as Edison had done, but commanding it to form a large, bright, and orderly spectrum. Edison devised safe and smokeless light for the American home; Rowland devised a way to transform the study of light in the physical laboratory. Like Edison, but far more obscure in the public landscape, he made his mark by invention. The result was a new means of knowing nature.

The device that Rowland envisioned and created became known as the concave diffraction grating, or simply the Rowland grating. (See Figure 1.) Like the prism, it spread white light into component colors, but unlike the prism, it operated by reflection rather than refraction, and made possible a relatively simple derivation of wavelengths from measurements of the spectrum, which was the step required for quantitative physics. Any diffraction grating brought this simplification, but existing ones required glass focusing and collimating lenses, which dimmed light by absorption. A concave grating, by distinction, would focus and diffract the light simultaneously by reflection, largely eliminating the need for lenses[47] and making the spectrum much brighter.

Because the analysis of light promised knowledge of the ultimate, still-mysterious constituents of the matter that produced light, and also

---

[45]Rowland and George F. Barker, "On the Efficiency of Edison's Electric Light" (1880), in *PPR*, 200-203.

[46]Remsen, "Henry Augustus Rowland," in *Leading American Men of Science*, 416.

[47]A condensing lens was still often used to concentrate more light on the concave instrument.

FIGURE 1. *Concave diffraction grating ruled at Johns Hopkins University. Rules area 5.2 centimeters by 8.1 centimeters. 14,438 grooves per inch. Courtesy National Museum of American History, negative #69718.*

because it promised knowledge of the elements composing the Sun and stars, any breakthrough in spectroscopic technique greatly enhanced the prospect of obtaining an understanding of the composition of matter on Earth, as well as the composition of the heavens.

Diffraction research was already an established, if unpopular, branch of spectroscopy. The study of light mostly relied on prisms, even though prismatic spectra yielded wavelengths only after laborious, nonlinear computations. Gratings were notoriously difficult to make or obtain, and even the best were little larger than a postage stamp. Limited size meant limited brightness of the resulting spectrum; hence the popularity of prisms. The more popular instrument dispersed light into its component colors by differential refraction: the glass in prisms possessed slightly different refractive indices for light of slightly differing wavelengths. So the prism bent light rays of one color more strongly than light rays of another color, fanning white light into a spectrum. But the scale of prismatic spectra was nonlinear: wavelengths could not simply be read from a linear scale, as they could from diffraction spectra.

By contrast, gratings relied on the principle of constructive interference. Their surfaces contained a great many fine, parallel, regularly spaced lines or grooves, which either reflected or transmitted light. Gratings concentrated light of different wavelengths at slightly different positions on a screen or photographic plate. This concentration occurred wherever the distance from one groove to a position on the screen, and the distance from an adjacent groove to that position, differed by one, two, or more whole wavelengths, corresponding to the "first-order," "second-order," and higher-order spectra, respectively.

Because the lines on a grating were evenly spaced, all combined to produce concentrations of light of one wavelength at the same positions. And because each succeeding color corresponded to a slightly different wavelength, each would appear at distinct positions at the screen; thus spectra were formed. In Rowland's view, the French physicist Augustin Fresnel had first stated the "true theory of diffraction" in 1816.[48] One Hopkins student later stated "the fundamental law of the [concave] grating" as: $Nl = w \sin Q$, where N is the order of the spectrum, l is the wavelength, w is the space between adjacent grooves on a grating, and Q is the angle between the incident light and a normal to the grating surface.[49]

The German optician Joseph von Fraunhofer was the first to make significant use of diffraction methods, about 1820, but the approach progressed very slowly through most of the nineteenth century. The limiting component of diffraction studies was always the gratings themselves. An American physician and amateur astronomer, John

---

[48]"Early notes as a student on the theory of light," 1, Box 20, HAR.

[49]Janet Tucker Howell, "The Fundamental Law of the Grating" (Ph.D. diss., Johns Hopkins, 1913). Reprinted in *Astrophysical Journal* (*ApJ*), 39 (1914): 230-242. Formula on 230.

William Draper, apparently was the first to photograph the solar spectrum by diffraction methods, in 1844.[50] Another American amateur, Lewis Rutherfurd, made and distributed a number of gratings, typically less than two inches across, in the 1870s and 1880s.[51] In Rowland's words, Rutherfurd gratings "showed more of the spectrum than had ever before been seen." He reported, "Many mechanics in this country and in France and Germany, have sought to equal Mr. Rutherfurd's gratings, but without success."[52]

Rowland's mechanical training from Rensselaer meant, however, that he did not have to wait for improvements on Rutherfurd; he achieved them himself. By careful machine design, and with the help of a machinist, Theodore Schneider, and Harvard astronomer William Rogers, who also experimented with grating manufacture, Rowland engineered a better approach. In his own department in the early 1880s, he found a way to surpass Rutherfurd's instruments in size and quality, at the same time that he employed concave surfaces for some of the gratings, for the property of focusing. After these innovations, spectra produced by the new gratings were bigger and brighter than any other diffraction spectra thus far. Consequently, physicists who used his instruments could claim to have a better grasp of whatever spectra they were studying and measuring. And like J.W. Draper and Rowland, they could apply another relatively new technology, photography, to record the spectra created.

Both concave and flat gratings from Baltimore became prized possessions in laboratories far and wide, but at Johns Hopkins a different kind of engineering also took place that, combined with the new instrument, synergetically enhanced the intensive analysis of light. Although Johns Hopkins was neither the first American institution to call itself a university, nor the first to award the Ph.D., it was the first ambitiously and systematically to cultivate professional, graduate training on the model of German universities. (In fact, graduate students outnumbered undergraduates at Hopkins until 1890.)[53] Even Harvard president Charles Eliot acknowledged a debt to Johns Hopkins for

> the creation of a graduate school, which has not only been itself a strong and potent school, but which has lifted every other university in the country in its departments of arts and sciences.

---

[50]Donald Fleming, "Draper, John William," Vol. 4 (1971), *DSB*, 183.

[51]Deborah Jean Warner, "Lewis M. Rutherfurd: Pioneer Astronomical Photographer and Spectroscopist," *Technology and Culture*, 12 (1971): 211-213.

[52]"Preliminary Notice of the Results Accomplished in the Manufacture and Theory of Gratings for Optical Purposes" (1882), in *PPR*, 487.

[53]Roger L. Geiger, *To Advance Knowledge: The Growth of American Research Universities, 1900-1940* (New York: Oxford University Press, 1986), 8.

I want to testify that the graduate school of Harvard University, started feebly in 1870 and 1871, did not thrive, until the example of Johns Hopkins forced our faculty to put their strength into the development of our instruction for graduates.[54]

Thus Johns Hopkins was a prime site for the engineering of a relatively new social technology: research schools, here defined historiographically as "small groups of mature scientists pursuing a reasonably coherent programme of research side-by-side with advanced students in the same institutional context and engaging in direct, continuous social and intellectual interaction."[55] In the sciences, Johns Hopkins offered graduate instruction in chemistry, mathematics, and biology, as well as physics. Existing historical studies make it clear that research schools were formed in both chemistry and mathematics;[56] a school probably existed in biology, as well.[57] In physics, two divisions of a research school were formed: one in electricity and magnetism, and one in the study of light, both directed by Rowland. The various schools of Johns Hopkins produced scientists in numbers great enough that, by the 1890s, "these made-in-America scholars were carrying the Hopkins spirit into all the major universities of the country."[58]

Through 1882, the year of optical invention, the physics department awarded only three Ph.D.s: one for work on the distribution of heat in the spectrum, one on electromagnetism (to Edwin Hall), and one on the mechanical equivalent of heat.[59] Clearly, the department was only getting started, six years after it had begun. The new instrument, the concave diffraction grating, changed that, as the inventor engineered apparatus, methods, and a research program to set work in motion.

---

[54]In *Celebration of the Twent0y-fifth Anniversary of the Founding of the University and Inauguration of Ira Remsen, LL.D. As President of the University* (Baltimore: The Johns Hopkins University Press, 1902), 105.

[55]Gerald L. Geison, "Scientific Change, Emerging Specialties, and Research Schools," *History of Science*, 19 (1981): 23. This definition has been pronounced "now-canonical." Servos, "Research Schools and Their Histories," *Osiris* 8 (1993): 13. (Volume titled *Research Schools: Historical Reappraisals*, eds. Geison and Frederic L. Holmes.)

[56]On chemistry, see Owen Hannaway, "The German Model of Chemical Education in America: Ira Remsen at Johns Hopkins (1876-1913)," *Ambix*, 23 (1976): 145-164. On mathematics, see Karen Hunger Parshall, "America's First School of Mathematical Research: James Joseph Sylvester at The Johns Hopkins University 1876-1883," *Archive for History of Exact Sciences*, 38 (1988): 153-196.

[57]In biology and physiology, H. Newell Martin came to Johns Hopkins from Cambridge University in England, where he had participated in the research school of Michael Foster. Thus Hopkins became "the leading training ground for the next generation of American biologists and physiologists." Geison, *Michael Foster and the Cambridge School of Physiology* (Princeton: Princeton University Press, 1978), 141-142.

[58]Geiger, *To Advance Knowledge*, 8.

[59]"Doctors' Dissertations, 1876-1926," *The Johns Hopkins UniversityPress Circular* 45, Whole Number 373 (1926): 41.

The first major studies, and the most important single use of the new instrument concerned the solar spectrum, which remained central to the school of light for at least twenty years. (The other major activity concerned the analysis of light from chemical elements in the laboratory.) The inventor first proved out the invention himself by taking successively more refined solar spectra. In 1883, he presented some of the first results at a meeting of the National Academy of Sciences in New Haven. A newspaper reported that he "astounded and amazed his scientific audience with a photograph of the visible solar spectrum in ten sheets, making a total length of sixty feet."[60] Rowland subsequently made photographic maps, with angstrom units marked, available for purchase, in sections three feet long. In the process, he publicly minimized the human agency involved in capturing the spectrum. A notice for the spectral maps in 1889 asserted, "The photograph is the work of the sunlight itself and the user of this map has the solar spectrum itself before him, and not a distorted drawing full of errors of wave length and of intensity."[61] The paradoxical effect of the physicist's ingenious engineering was to construct something entirely natural, although nature here bore the imprimatur of his university.

In the 1880s, having taken images of the solar spectrum and found them satisfactory, Rowland began turning out three products in quantity: the gratings, for use at many locations, in both solar and laboratory studies; copies of the solar maps; and physicists expert in the use of the gratings and of other spectroscopic instruments. Together with his own further investigations, these products carried with them a distinctive approach to the study of nature.

Spectroscopy is important to the historiography of physics in the United States because it was, numerically, the dominant specialty in the late nineteenth and early twentieth centuries, the period in which a modern profession of physics emerged. By 1900, in numbers, American physics was not a small power. It ranked with Germany and Britain as one of the three nations having the greatest number of academic physicists, although Americans were considerably less productive than their European counterparts, if published papers provide the measure.[62] However, in the case of Rowland, productivity cannot be measured by published pages alone.

---

[60]*Baltimore American*, 18 November 1883, in Rowland scrapbook, Manuscript Collection 1045, American Philosophical Society, Philadelphia.

[61]"Photographic Map of the Normal Solar Spectrum," *Johns Hopkins University Press Circulars*, no. 73 (1889), 80, in Box 35, HAR.

[62]Paul Forman, John L. Heilbron, and Spencer Weart, "Physics *circa* 1900: Personnel, Funding, and Productivity of the Academic Establishments," *Historical Studies in the Physical Sciences*, 5 (1975): 6, 12.

Spencer Weart's historical examination of the physics profession prior to World War II notes that "Americans had made a specialty of manipulating light" since the nineteenth century. As late as 1935, the most common specialty for which American physics departments claimed expertise was spectroscopy.[63] Robert Rosenberg has shown that the needs of industry for knowledge of electricity and magnetism also fueled the growth of academic physics departments in the 1880s.[64] But electricity and magnetism did not crowd out spectroscopy, and were sometimes taught in new programs or departments in electrical engineering.

The quality of research training specifically at Johns Hopkins was evinced impressively, shortly after Rowland's time, by polling. Of eighty-five American physicists elected by leaders in their field for special recognition in 1903 (in all branches of physics), twenty-five received their doctorates from Johns Hopkins. Cornell ranked next with thirteen, followed by a German university, Berlin, with twelve. Columbia was a distant fourth, with six. The Johns Hopkins count was a significant measure of one department's power to promote research physics. It also reflected the larger strength of Johns Hopkins in graduate education.[65]

The fecundity of the Hopkins school was not limited to Rowland's lifetime. During his twenty-five years at Johns Hopkins (1876-1901), the department produced forty-five Ph.D.s, of whom sixteen specialized in the study of spectra; another twenty studied electricity and magnetism. There were actually 165 graduate students altogether, working in various fields in the department between 1879 and 1901, but many did not take degrees. During the second quarter-century (1902-1926), the department continued to foster spectroscopy, producing forty-four Ph.D.s who studied spectra, out of eighty-four recipients in total. Thus the half-century total in spectroscopy was sixty.[66]

For those who studied light, the most basic phenomenon was the spectral line, which was essentially an image, in one very specific color, of the narrow slit through which light entered a spectroscope. Not all spectra exhibited spectral lines; some were continuous. In

---

[63]Weart, "The Physiczs Business in America, 1919-1940: A Statistical Reconnaissance," in *The Sciences in the American Context: New Perspectives*, ed. Nathan Reingold (Washington, DC: Smithsonian Institution Press, 1979), 300-301.

[64]Robert Rosenberg, "Academic Physics and the Origins of Electrical Engineering in America" (Ph.D. diss., Johns Hopkins, 1990), 5, 271.

[65]Stephen Sargent Visher, *Scientists Starred, 1903-1943, in 'American Men of Science'* (New York: Arno Press, 1975), 3, 291 table, 278 table. (First published by The Johns Hopkins University Press, 1947.)

[66]"Doctors' Dissertations, 1876-1926," 41-47. On the total of 165 students between 1879 and 1901, see Robert Kargon, "Henry Rowland and the Physics Discipline in America," *Vistas in Astronomy*, 29 (1986): 131.

some spectra of hot bodies, notably the famous spectrum of the Sun which Isaac Newton obtained in the seventeenth century, no spectral lines were apparent. The blending of one color into the next, then into the next, was insensible and continuous from the extreme red end of the spectrum to the extreme violet end. The length of any spectrum depended on the power of a spectroscope to separate light into colors, or the dispersion. The width of the spectrum depended upon the length of the entrance slit. There were practical limits to the dispersive power of spectroscopes, and so to the magnification of the spectrum, but Rowland expanded them.

Spectral lines were a new phenomenon in the nineteenth century. When testing optical glasses in the 1810s, Joseph von Fraunhofer obtained the familiar solar spectrum, but on close examination found that it was not completely continuous. Rather, fine, dark lines crossed it, interrupting the gradations of color that had previously appeared seamless. Newton had not observed this. (Fraunhofer utilized prisms initially, and eventually also gratings.) In other words, Fraunhofer found that, at certain wavelengths, the bright image of the slit was absent, that the Sun furnished no light of those extremely specific colors. The Bavarian optician noted 574 lines in total.[67]

The study of artificial light sources in the laboratory revealed an opposite type of spectral line—bright. Spectra of bright gases mostly consisted of darkness, interrupted at very discrete places by bright lines, meaning that the gases furnished light only at very specific wavelengths. But whether bright or dark, all spectral lines were essentially images of the opening through which light entered a spectroscope. The measurable properties of the lines basically consisted of wavelength and brightness (or darkness), although the exact shape of a line might vary, and might be noted, too. The name for each line was its wavelength. Indeed, in tabular form, a line and its wavelength became much the same thing.

Not all spectra exhibited lines. Newton's solar spectra did not show lines at least partly because the resolving power of his spectroscope (a prism) was low. Yet even when viewed with refined apparatus, the spectra of incandescent solids, such as the filament in Edison's electric lamp, did not exhibit spectral lines. These spectra were continuous, and in general commanded less attention from spectroscopists of the Hopkins school than line spectra, with their discrete,

---

[67]Fraunhofer was not the first to see dark lines in the solar spectrum; William Wollaston noticed lines in 1802. But Fraunhofer recorded a large number of them systematically. William McGucken, *Nineteenth-Century Spectroscopy: Development of the Understanding of Spectra, 1802-1897* (Baltimore: The Johns Hopkins University Press, 1969), 2-4.

measurable wavelengths. Spectra of incandescent solids fell, perhaps paradoxically, under the category of blackbody radiation.[68]

Typically, the historiography of light at the start of the twentieth century emphasizes two historical discontinuities. One involved the conceptual shift to the idea of quantized energy levels in the atom. The second usually-noted leap in the new century was the special theory of relativity, in which one postulate held that the speed of light in empty space was independent of the speed of the body emitting that light.[69] Together, the conceptual innovations of the quantum and of relativity are often taken to represent much of the revolutionary character of the physics of light at the opening of the twentieth century.

Yet the arrival of a new century did not invalidate all previous physics. For the most part, the transition from 1899 to 1900 caused no upheaval; it did not require a radical break with the past, despite the historiographical convenience which would accrue to the modern historian of science, if it had done so. Many fields continued work unperturbed at this milestone; one of them was the study of spectra, with roots firmly fixed in the earlier century. Wherever physicists studied the spectra of gases, or the spectra of celestial bodies, they studied physical discontinuities, the spectral lines demarcating the spectrum. (Studies of the continuous or blackbody spectrum plotted relatively smooth curves of intensity versus wavelength.) This study of discontinuities reached from the old century into the new, and the implementation and use of Rowland's instrument belonged to both centuries, as did the quest for a rational understanding of spectra.[70]

Spectroscopy held great significance for understanding the heavens. The far-reaching implications had become apparent from 1859, when physicist Gustav Kirchhoff, working in Heidelberg with chemist Robert Bunsen, concluded that specific spectral lines always signified the presence of specific chemical elements. Bright lines resulted from emission by elements; the same lines, if dark against a bright continuous spectrum, indicated absorption by the same elements. Further, when lines in the spectra of the Sun and stars had the same wavelengths as lines obtained in the laboratory, they indicated the presence of the same elements. The heavens no longer consisted of ineffable ce-

---

[68]Three Americans who did study blackbody spectra were Charles E. Mendenhall and Frederick A. Saunders, both of Rowland's school, and Samuel Langley, who used a Rowland grating. Mendenhall and Saunders, "The Radiation of a Black Body" (Ph.D. diss., Johns Hopkins, 1901). Langley, "Sunlight and Skylight at High Altitudes" [report of address], *Nature*, 26 (1882): 587. Also, in the Hopkins school, W.W. Jacques (Ph.D. 1879) and H.F. Reid (Ph.D. 1885) carried out research on the distribution of energy in spectra.

[69]Albert Einstein, "On the Electrodynamics of Moving Bodies," in *The Principle of Relativity*, trans. W. Perrett and G.B. Jeffery (New York: Dover Publications, 1952), 38.

[70]The older history of the quest is examined in McGucken, *Nineteenth-Century Spectroscopy.*

lestial matter but of familiar elements. Where only the position, brightness, and approximate color of a star were known previously, astronomers could now in principle deduce the presence of specific chemical elements.[71]

This was still the problem situation when Rowland entered the field in the 1880s. His new instrument, the concave diffraction grating, rapidly became the centerpiece of a school of spectroscopy at his department, and his study of the solar spectrum became the largest single research project. There is no evidence that Rowland, prior to inventing the instrument, planned to direct a systematic study of light; after the invention, he could hardly fail to do so. The analysis of light promised clues to the nature of matter, as well as the chemical composition of the Sun and stars. Rowland's mastery of light greatly differed from the conceptual conquests of Planck and Einstein; unlike them, that he realized it with screws, diamonds, pulleys, metal plates, water motors, wooden beams, wax, strings, mirrors—and engineering. (See Chapter 2.)

The insight of Kirchhoff and Bunsen suggested an almost inexhaustible program of research: the cross-comparison of laboratory and astronomical spectra. Although, over time, many took up the task, no one in the nineteenth century completed a more thorough study of the solar spectrum than Rowland, starting a little more than two decades after the unifying principle was articulated at Heidelberg. Rowland published successive solar wavelength tables, culminating in his final tables (which he still titled "preliminary") in the 1890s, in which he catalogued some 20,000 solar absorption lines, each expressed to the nearest thousandth of an angstrom unit, which unit represented 10^{-10} meters.[72] (Between Fraunhofer and Rowland, others had captured the solar spectrum at increasingly large scales. Their number included Draper, Anders Ångström, and Rutherfurd.)

Because of the vast implications of Kirchhoff's ideas and Rowland's tables, there could be little cause for idleness among students working in a well-equipped spectroscopic laboratory. The master mainly kept for a paid assistant, Lewis Jewell, the immense task of

---

[71]As in the case of Fraunhofer and the solar absorption lines, partial insights were achieved prior to Kirchhoff and Bunsen, who were the first to apply the principle aggressively and systematically. The forerunners included William Thomson, George Stokes, and Léon Foucault. On these three, and on the Heidelberg two, see McGucken, 22-24, 27-34. On Thomson and Stokes, see also Smith and Wise, *Energy and Empire*, 397-398.

[72]"Preliminary Table of Solar Spectrum Wave-Lengths. I." *Astrophysical Journal* 1 (1895): 29-31, followed by lengthy wavelength tables, in installments in this and the two succeeding volumes of the journal. The tables were also published separately, bound in one volume, in Rowland, *A Preliminary Table of Solar Spectrum Wave-Lengths* (Chicago: The University of Chicago Press, 1898).

measuring all lines in the solar spectrum. To ground an attempted understanding of the solar spectrum[73] by 1895, Rowland and others at Hopkins had also photographed the spectrum of every then-known terrestrial element except one. Further, more mundane laboratory studies of light from the terrestrial elements promised almost limitless investigations for students as well as professors, especially when conditions, such as temperature and pressure, under which luminous matter radiated were varied. The master reference was always Rowland's solar spectrum. The Hopkins school assumed that each solar line should correspond to a laboratory line.[74] Rowland hoped to see all the lines of the solar spectrum identified with lines observed from laboratory samples of the elements,[75] but the goal was not achieved in his lifetime, in part because his instrument made so many solar spectral lines newly visible.

The new instrument enlarged spectroscopy by enlarging the spectrum, revealing that many previously known spectral lines were in fact themselves composed of multiple, finer lines. Each of the components, noted by wavelength and intensity, then became a discrete fact. In this sense, the engineering physicist made the field twenty times larger than it had been, as measured by the number of wavelengths in his solar tables. Anders Ångström, who produced the most authoritative map of the solar spectrum before Rowland, observed and drew 1,000 lines.[76] As noted above, the inventor of the concave grating photographically captured 20,000.[77]

Rowland's establishment of a research school differentiated him from previous analysts of solar light. Fraunhofer, in his day, achieved insights that were only re-established many decades later, in part by the Hopkins school.[78] Because Fraunhofer left behind no students versed in the technical requirements of diffraction spectroscopy, his

---

[73]"Preliminary Table," 29, 30. Rowland could not obtain a specimen of gallium. 29.

[74]Klaus Hentschel has explored the implications of the essential postulate in spectroscopy that lasted from the 1860s until at least the 1890s. He wrote, "The assumption of the precise coincidence of solar and terrestrial spectral lines soon became the mainstay of the then rapidly evolving science of spectroscopy." Hentschel, "The Discovery of the Redshift of Solar Fraunhofer Lines by Rowland and Jewell in Baltimore around 1890," *Historical Studies in the Physical and Biological Sciences*, 23, part 2 (1993): 257.

[75]"Preliminary Table," 29.

[76]J.B. Hearnshaw, *The Analysis of Starlight: One Hundred and Fifty Years of Astronomical Spectroscopy* (New York: Cambridge University Press, 1990), 4.

[77]"Preliminary Table."

[78]See Joseph S. Ames, "Preface," in *Prismatic and Diffraction Spectra: Memoirs by Joseph von Fraunhofer*, ed. Ames (New York: Harper & Brothers, 1898), v-vi. Reprinted in *The Wave Theory of Light and Spectra*, ed. I. Bernard Cohen (New York: Arno Press, 1981). The insights included the effect of periodic errors in gratings, and the effect of groove shape. Ames was himself a product of Rowland's school (Ph.D. 1890).

knowledge survived only in printed form, not in the work of living physicists. Likewise, no schools emerged around Draper or Rutherfurd. Rowland, by contrast, enrolled many students in his spectroscopic laboratory, where they witnessed, and not simply read about, his laboratory techniques, and could follow his lead. (The engine that produced the gratings remained foreign even to most of his students.) Many students, both during and after the master's career, utilized Rowland's "big spectroscope,"[79] which he built under and around a concave diffraction grating. A number of them used it for their doctoral studies. For reference, they also used the solar tables.

As they measured wavelength values, students of the Hopkins school, both during and after Rowland's time, may also have absorbed some of the master's professional values, which were sometimes reflected in their later work. First, interest in the celestial-terrestrial comparison, especially the program of correlating solar and elemental lines, was fundamental to the school, and permeated the work of virtually all who analyzed light. Physicists trained at Hopkins sometimes worked in astronomy, especially on instruments. Indeed, the most common place for Hopkins students of spectroscopy to publish their doctoral research was the *Astrophysical Journal*, which began publication in 1895.[80] The comparability of earthly and heavenly light was a fundamental tenet of the editors.

In the long run, diffraction instruments produced at Johns Hopkins greatly enhanced the spectroscopy of the stars, decades after the founder released his tables on the closest star, the Sun. Through the comparison of astronomical and laboratory light, Rowland led his students and others to see on Earth the material of the stars. (Had he been present, Rowland's grandfather the minister might have endorsed the dual approach, for what it would reveal. He once said of the heavens: "None but a being of infinite wisdom could have planned the vast machines—could have hung them in empty space and balanced worlds and systems; and caused beauty, harmony and order to shine in all.")[81]

Another of the master's values later apparent in the careers of his apprentices was internationalism, the desire and readiness to cross national boundaries in the pursuit of physics. As the Baltimore school, and later other American schools, emerged, graduate study in Europe, especially Germany, gradually became less essential to American students. But among the first Hopkins students who became prominent, virtually all of them studied for a short time in Germany, then the

---

[79]The designation "big spectroscope" appears in the jottings of student Henry Crew, 1886 Diary, 15 February, HC.

[80]"Doctors' Dissertations, 1876-1926."

[81]Henry A. Rowland [I], *A Discourse*, 8.

wellspring of physics, as well as Baltimore. The science of physics was unquestionably further advanced in Germany in the late nineteenth century. Even in the 1920s, when the physics profession was well established in the United States, Hopkins-trained investigators of light still often looked abroad for conceptual leadership. Yet concepts were not the only aspect of physics open to international conciliation; wavelengths were another. Various physicists and astronomers, both within and outside the Hopkins group, believed that attempts to establish standard wavelength values for solar and elemental spectra should be international in organization. Rowland hoped to have the last word with his solar tables, but instead they became one resource, among many resources of lesser scope, in international attempts to agree on specific wavelengths as universal benchmarks.

Third, a mechanical facility appeared in an important few, but by no means all, of the physicists produced at Hopkins. Rowland's mechanical ingenuity did not necessarily reappear in the work of the physicists he trained. The founder felt that he understood light once he could control it; his students benefitted from the means of control. Although tinkering with, or adjusting, experimental apparatus was always necessary, the students were not all trained as engineers, as Rowland was. Rather, they were beneficiaries of his laboratory engineering; when they began their researches, the technology requisite for knowledge production was already at hand. Thus, while ingenuity in experimental apparatus was a distinguishing mark of Hopkins physics, it did not necessarily recur in the later work of all who passed through there.

Two physicists who did exhibit mechanical skill were John Anderson (Ph.D. 1907) and Robert Williams Wood, son of Robert Williams Wood, a physician. Both physicists eventually oversaw the operation of Hopkins ruling engines, the delicate and temperamental machines that turned out Rowland gratings. Even when a certain veil of secrecy about the engines was lifted, when mechanical drawings were published in 1902,[82] few if any elsewhere could immediately reproduce the engines. Thus the skill necessary in making high-quality diffraction instruments continued to distinguish at least two of the many physicists who passed through the Hopkins laboratory. Their products were still highly prized, well into the twentieth century. The work of Rowland, Anderson, and Wood in furnishing laboratories in many parts of the world with high-dispersion instruments was a major component of the Hopkins legacy. The international community of spectroscopists would have received great benefit from

---

[82]Between pages 698 and 699, *PPR.*

instrument production alone at Hopkins, even without the research school. The combination of the two helped to launch the American physics profession.

Fourth and finally, the research school for the study of light can be seen as an expression of Rowland's vision for pure science in America, and as one means of fostering and conserving a place for science, especially physics, in American society. Materially an innovator, he was conceptually a conservative. In the year following his invention, the physicist delivered his most famous address, "A Plea for Pure Science," to the American Association for the Advancement of Science in Minneapolis.[83] He argued that a "grand laboratory" should be created, one which "does not exist in the world, at present, for the study of physics." (The closest approximation, not mentioned, was probably Helmholtz's laboratory.) The most important component of the laboratory would be its director, who would create a plan of research and implement it. More laboratories would ensue. Rowland predicted, "After one has been successfully started, others could follow; for imitation requires little brains."[84]

David Hounshell has asserted that Rowland's address "was in large part a response to the scientific, technical, and social milieu of the time—and specifically, to Edison's activity in electric lighting."[85] Indeed, the speaker, at least in this address, heaped scorn on commercial applications:

> . . . it is not an uncommon thing, especially in American newspapers, to have the *applications* of science confounded with pure science; and some obscure American who steals the ideas of some great mind of the past, and enriches himself by the application of the same to domestic uses, is often lauded above the great originator of the idea, who might have worked out hundreds of such applications, had his mind possessed the necessary element of vulgarity.[86]

The physicist did not mention the inventor by name, but the contrast between his view of science and Edison's was implicit. He stated simply, "The proper course of one in my position is to consider what must be done to create a science of physics in this country, rather than to call telegraphs, electric lights, and such conveniences, by the name of science."[87] Edison manufactured profitable and artificial light; Rowland

---

[83]"A Plea for Pure Science," 593-613.

[84]Ibid., 607-608.

[85]David A. Hounshell, "Edison and the Pure Science Ideal in 19th-Century America," *Science*, 207 (1980): 616.

[86]"A Plea for Pure Science," 594. Emphasis in original.

[87]Ibid., 594.

analyzed pure and natural light. (The physicist did not always scorn profit. His ventures into commerce included consulting for a hydroelectric power project[88] and the development of an "octoplex printing telegraph."[89] Three years before the "Plea," he even planned to seek a joint patent with Edison for an electromagnetic device.)[90]

The physicist must have looked on with dismay when Edison provided financial backing in 1880 for a new journal entitled *Science*. It ceased publication after two years, but the title re-emerged in 1883 with the financial backing of the inventor of the telephone, Alexander Graham Bell. (Publication ceased once again in 1894, and the present journal arose with the support of the American Association for the Advancement of Science.)[91] Rowland fought to differentiate science from commercial technical innovation.

Although Rowland's invention for the analysis of light eventually brought about dramatic change in physics, the physicist had little inclination to upset the edifice of science imported from Europe. Even where conflict existed, as between James Clerk Maxwell's electromagnetic theory of light and William Thomson's elastic-solid theory of light, the American preferred peaceful coexistence: he taught both theories.[92] Conceptually, although Rowland admired Galileo, Copernicus, and Luther, the Hopkins-school physicists were scientific conservatives, not revolutionaries. They retained an all-pervasive ether as the accepted medium for the propagation of light waves. Until the 1920s, they, like others, believed that the Sun possessed much the same composition as the Earth, albeit at a vastly different temperature. One of Rowland's successors, Joseph Ames, believed that the general theory of relativity, after it appeared in 1915, was scarcely a theory at all.[93] Mechanical models were the desired mode of explanation.

Although Hopkins physicists, and more importantly Hopkins instruments, flooded the spectroscopic community with wavelength

---

[88]Rowland consulted for the Niagara Cataract Construction Company, and later sued them for underpayment. Stanley M. Guralnick, "The American Scientist in Higher Education, 1820-1910," in *The Sciences in the American Context*, ed. Nathan Reingold (Washington: Smithsonian Institution Press, 1979), 132.

[89]Remsen, "Henry Augustus Rowland," 425.

[90]Rowland to Gilman, 1 August 1880, Daniel Coit Gilman Papers (DCG), Ms. 1, Special Collections Department, Milton S. Eisenhower Library, Johns Hopkins University Press.

[91]The emergence of the journal is recounted in Sally Gregory Kohlstedt, "*Science*: The Struggle for Survival, 1880 to 1894," *Science*, 209 (1980): 33-42.

[92]This is reflected in the notes of a student. Joseph Ames, "Rowland's Lectures on Light," 1888, Joseph S. Ames Papers (JSA), Ms. 61, Special Collections, Milton S. Eisenhower Library, Johns Hopkins University Press.

[93]He wrote, "it is not correct to say that Einstein has proposed a 'theory' of anything, in the sense of giving a picture of any mechanism." Ames, "Einstein's Principle of Relativity and Its Bearing Upon Physics," *Journal of the Franklin Institute*, 191 (1921): 2.

data, the school was intent on conserving and transmitting extant physics, rather than on overturning it. It also, through the agency of a few men, conserved the ruling engines, by maintaining a working knowledge of them, and even by improving them. Rowland was so concerned with making a space for physics in America that he had little time to quarrel with the conceptual status quo as imported from Europe, except to compare experimental results from his laboratory with those from abroad.

In the analysis of light, the essential insights being conserved were the principles of Kirchhoff and Bunsen. Each identification of a solar line proved, it was thought, that Earth and Sun were much alike, and thus the composition of the cosmos was comprehensible. It was only in the hands of other physicists, and astronomers, and only after some decades, that the welter of orderly and carefully-measured shadow lines in dispersed sunlight began to mean something different.

If Rowland was the founding engineer of his own school, Henry Crew was, eventually, its philosopher. Crew received his Ph.D. from Hopkins in 1887, and spent most of his career teaching at Northwestern University. When he wished, in the 1920s, to provide "an excellent illustration of good scientific method," he found it in the discipline of spectroscopy, and in the fundamental and long-familiar insights of Kirchhoff and Bunsen into the significance of spectral lines. Sound method, exemplified by the German physicist and chemist, involved, he explained, five steps:

(i)   The establishment of the experimental facts.
(ii)   The proper pigeon-holing of the facts; i.e., their allocation with other facts of essentially the same kind.
(iii) The impersonalization of the facts; i.e., elimination of the observer's subjective self.
(iv) The formulation of the facts; i.e., their expression, if possible, in the form of a physical law.
(v)   The prediction of new facts.[94]

The prediction implicit in Rowland's photographic maps of the solar spectrum in the 1880s and 1890s was that for every celestial fact, every absorption line in the solar spectrum, there would be a corresponding terrestrial fact, an emission line from an element isolated in the laboratory and made to shine by flame or electric arc or spark. If the school followed a conceptual paradigm, it was that of Kirchhoff and Bunsen. The personal paradigm was Rowland himself, the active inquirer.

---

[94]Crew, *The Rise of Modern Physics* (Baltimore: The Williams & Wilkins Co., 1928), 306.

In Crew's formulation, a physicist started with a small number of facts, performed his operations, and finished with a larger number. In spectroscopy, the irreducible facts were wavelengths of spectral lines. Rowland, the physicists of his school, and the physicists who used his instruments, produced thousands of such facts. The principal carryover into the twentieth century was thus a welter of numbers calling for mathematical interrelationship. But the purpose of formulas, in Crew's philosophy, was simply to suggest paths to discovery of yet more facts.

These facts mattered because they were representations of nature. In education, Rowland believed, "It is necessary that we have some standard of absolute truth: that we bring the mind in direct contact with it and let it be convinced of its errors again and again." The student "must be brought face to face with nature."[95] The solar spectrum was one manifestation of nature, and the study of that spectrum, in conjunction with laboratory spectra, promised better access to the ultimate source of truth, nature. Because of this approach to truth, physics at Hopkins suggested a higher calling. As Daniel Kevles has observed, the physicist's idea of pure science, expressed in 1883, carried "the connotation of high virtue."[96] Physics was an altogether improving endeavor, because it concerned nature. Rowland bound physics, too, to an older American dream:

> For pure science is the pioneer who must not hover about cities and civilized countries, but must strike into unknown forests, and climb the hitherto inaccessible mountains which lead to and command a view of the promised land,—the land which science promises us in the future; which shall not only flow with milk and honey, but shall give us a better and more glorious idea of this wonderful universe.[97]

The western frontier, where nature and European man met, was still a special feature of American life in 1883, and Rowland here essentially described a parallel between westward migration and science. Nature ennobled the scientist, as it might elevate Americans in westward migration.[98]

Yet the Hopkins school did not study forests and mountains. Sunlight was the only thing it studied that was present in the natural landscape. Otherwise it studied the light of chemical elements, which were

---

[95]"The Physical Laboratory," 615-616.

[96]Kevles, *The Physicists: The History of a Scientific Community in Modern America* (Cambridge: Harvard University Press, 1987), 45.

[97]"A Plea for Pure Science," 608.

[98]Historian Frederick Jackson Turner, who studied history at Johns Hopkins in 1888-1889, made the classic argument that the western frontier defined an American national character. On the relation between science, specifically the life sciences, and Turner's work, see William Coleman, "Science and Symbol in the Turner Frontier Hypothesis," *American Historical Review*, 72 (1966): 22-49.

taken to be natural, but which were far more purified in the laboratory than matter existing in nature. Because the study of light was so inextricably bound up with the development of instruments, the progress of science in the Hopkins school followed close on the heels of progress in machining screws, utilizing diamond ruling points, and eliminating tremors from the laboratory. Such progress undergirded the conceptual unification of the heavens and the Earth. If the faithful listening to Rowland's grandfather in 1800 was required to "have a heavenly temper formed within,"[99] so was the matter that the Hopkins school studied in the laboratory after 1882; it was made to shine near the temperature of the stars. The program of reconciling celestial and terrestrial light became, for a time, an institution of American physics.

---

[99]Henry A. Rowland [I], *A Discourse*, 6.

2

Analysis Materialized

Neither the naked hand nor the understanding left to itself
can effect much. It is by instruments and helps that the
work is done, which are as much wanted for the
understanding as for the hand. And as the instruments of
the hand either give motion or guide it, so the instruments
of the mind supply either suggestions for the
understanding or cautions.
- *Francis Bacon, 1620*[1]

For Rowland, the spectra he produced were phenomena that nearly, or
ideally, represented nature itself. But they also embodied his own
agency and skill, and in this sense the apprehension of the natural,
and the founding of a research school, required great ingenuity. The
spectra that Rowland, his students, and the far-flung owners of Hopkins-made gratings obtained, were to a great extent accepted as natural manifestations of the chemical elements or of the sun, but they
were also products of elaborate and intricate machining in the base-
ment of the Johns Hopkins physics department. Starting from knowledge of screws and fine mechanical motion, the Baltimore physicist
reached into various laboratories in America and Europe, where his
instruments, showing greater control than previous devices, artificially
produced natural, normal spectra. Physicists of many cities and nations, as well as Johns Hopkins students, benefitted from an ability to
divide and rule the many colors of light, based on Rowland's ability
to divide and rule many surfaces of speculum metal.

---

[1]Francis Bacon, *The New Organon and Related Writings*, ed. Fulton H. Anderson (Indianapolis:
Bobbs-Merrill Educational Publishing, 1981), 39.

Those who watched the Baltimore master at work, especially his students, could well see the agency of the physicist in producing natural spectra. One graduate student in chemistry, Robert W. Wood, visited in the physics laboratory but did not take a degree from Hopkins in either chemistry or physics. However, he eventually became one of Rowland's successors and a steward of the ruling engines that produced the gratings. When he was a professor of experimental physics at Hopkins, Wood described the human agency involved in creating any spectrum, whether by grating or by prism. Although he recounted Newton's demonstration that the various colors of light mixed together are present in white light, he also wrote, "Our present idea regarding the action of the prism more nearly resembles the idea held previous to Newton's classical experiments: we now believe that the prism actually manufactures the colored light." So did the grating. Although he allowed that Newton's understanding could still inform much physical explanation, Wood questioned its applicability to all spectrum research: "How are we to regard a vibration which is made up of an infinite number of regular trains of waves, each, however, of different wave-length?"[2]

For Wood, the idea of the heterogeneous composition of white light might owe too much to appearances; it may have misconstrued the action of the prism, because "the dispersion by prisms and gratings can be accounted for without assuming the presence in the light of any periodicity whatever. Up to the present time no experiment has been devised capable of proving or disproving the presence in white light of regular wave-trains." The conventional attribution of the colors of the spectrum to an intrinsic mixture in white light ignored the agency of instruments in producing the spectrum: "It is easy to see how periodicity can be manufactured by a grating or prism." Fourier analysis could show a pulse of light to be an apparently irregular wave, resulting from many simple sine waves. In this sense "the spectroscope will sort out the Fourier components into periodic trains of waves, just as if these wave-trains were really present in the incident light."[3] This analysis took place at a material rather than an intellectual level, and in this sense a spectroscope was analysis materialized.

In the case of diffraction gratings, Wood introduced an analogy with sound, with a commonplace effect which anyone could perform. He first described the acoustical effect: "If a sudden sharp noise such as is made by clapping the hands together is reflected from a high flight of steps, the sound comes back to us as a musical note; in other

---

[2]Robert W. Wood, *Physical Optics* (New York: The Macmillan Co., 1911; reprinted 1924), 1, 648.

[3]Ibid., 649, 653-654.

words, the steps impress the element of periodicity upon the reflected disturbance, each step throwing off an echo-wave." The action of a grating on white light was, he explained, much the same.[4] (In fact a later form of grating, the echelon grating devised by Albert Michelson in 1898, consisted of a series of overlapping glass plates, configured at their edges much like a miniature staircase.)[5]

While either a prism or a grating could perform a physical Fourier analysis on light, a prism would still leave difficult computations to be performed, because the distance between spectral lines in a prismatic spectrum did not stand in direct proportion to the difference between the wavelengths of the lines. If plotted on a graph, the relation between distance of separation and the wavelength would form a curve rather than a straight line. For grating spectra, on the other hand, the separation of lines was directly proportional to the wavelength so that, if the wavelength of one line were known, simple arithmetic would yield the wavelengths of the other lines. Thus, the instrument minimized the work required from a physicist. A grating not only performed an analysis of the periodic components of light, it also normalized the spectrum. The use of a suitable measuring scale could even allow for direct reading of wavelengths. The grating ordered phenomena.

Rowland did not invent diffraction spectroscopy; he simply improved on a difficult art. The first person to use a diffraction grating was probably the American David Rittenhouse in the 1780s. His device comprised 50 hairs stretched between two fine screws, with adjacent hairs kept apart by the screw threads, at a density of 106 per inch. The whole grating was only one half-inch square, but it enabled Rittenhouse to view diffraction spectra. (Following Newton, he called the effect "inflection.") Rittenhouse concluded that, through diffraction studies, "It is probable that new and interesting discoveries may be made, respecting the properties of this wonderful substance, light, which animates all nature in the eyes of man, and perhaps above all things disposes him to acknowledge the Creator's bounty. But want of leisure obliges me to quit the subject for the present."[6]

The principal founder of diffraction spectroscopy, however, was Joseph von Fraunhofer, a Bavarian optician working in a secularized monastery, who published his diffraction work in the 1820s. Fraunhofer also employed fine screws as structural components in his gratings, but used wires rather than hairs stretched between. His wires

---

[4]Ibid., 656-657.

[5]C. Candler, *Practical Spectroscopy* (Glasgow: Hilger & Watts Ltd., 1949), 180.

[6]David Rittenhouse, "A Problem in Optics," *Transactions of the American Philosophical Society*, 2 (1786): 202-206.

measured two thousandths of a "Paris inch" in thickness, and were separated by about four thousandths. Fraunhofer had, in the previous decade, used a prism when he discovered the hundreds of absorption lines in the solar spectrum that today bear his name (the Fraunhofer lines), but he saw promise in diffraction methods, which displayed absorption lines of the same intensity and in the same order as lines obtained by prism, but with different spacing. The use of gratings enabled Fraunhofer to obtain the wavelengths of the absorption lines, rather than simply to note their relative order, as he had done with a prism.[7]

There could be many ways of making a grating: "It is entirely immaterial in a grating which is to be used for these experiments whether the threads out of which it consists are opaque, translucent, or transparent. A grating of glass fibres, for instance, produces these phenomena as well as one made out of metal wires." But Fraunhofer went beyond fibers, and created a machine that used a diamond point to rule lines on a glass surface. He produced reflection as well as transmission gratings. Much depended upon the exact shape of each diamond point. If it were sharp enough, the resulting lines would be so fine as to be nearly indistinguishable from each other, even under a powerful microscope. He believed that he had limited any possible errors in spacing to about one ten-thousandth of an inch. His most finely ruled gratings yielded spectra as large as those from large prisms.[8]

Although Fraunhofer's instruments and observations were, in the long term, portentous for physics and astronomy, his approach was phenomenological, and he drew few inferences from the spectra he observed. It was he who first assigned the letter *D* to one of the most prominent (double) lines in the solar spectrum, but he placed little interpretation upon it, except to observe that, in the spectra of some flames in the laboratory, "There is in the orange a bright line which is prominent above the rest of the spectrum, is double, and is at the same place where in sunlight the double line D is found." The deduction that sodium, which may produce these lines in a flame, is also present in the sun, postdated Fraunhofer. However, the Munich optician, sometimes working with prisms, did extend his studies from the sun to other celestial bodies. He found that the moon, Mars, and Venus all produced spectra containing some of the same dark lines as sunlight. Starlight to some extent resembled planetary light, but Fraunhofer found lines in the spectrum of Sirius unlike any lines from the planets.[9]

---

[7]Joseph von Fraunhofer, "New Modification of Light by the Mutual Influence and the Diffraction of the Rays, and the Laws of this Modification" (1821-1822), 21-23, in *The Wave Theory of Light and Spectra*.

[8]Fraunhofer, "Short Account of the Results of New Experiments on the Laws of Light, and Their Theory" (1823), 43-46, 54, in *The Wave Theory*.

[9]Ibid., 58-60.

In terms of Fraunhofer's own professional goals, the elucidation of celestial spectra was secondary to the perfection of optical apparatus. The solar absorption lines conveniently and exactly marked off different portions of the solar spectrum, helping Fraunhofer to appraise the properties of various optical glasses in controlling light of different colors. Fraunhofer described the difficulties accompanying his studies of light:

> It is greatly to be regretted that they can be repeated so seldom by any one, owing to the fact that they demand very large, and, in part, expensive apparatus, and also a great deal of time. The fact that the sky must be favorable makes one lose more time than would be believed, perhaps; which I feel all the more because the demands of business leave me only a few definite days in the month which are free for these investigations.[10]

Although the optical firm that employed Fraunhofer prospered, a few decades lapsed before other investigators, most notably Kirchhoff and Bunsen, took up the physical interpretation of spectral lines systematically.

Fraunhofer glimpsed problems and solutions that remained topical even one hundred years later. First was the principle of plane gratings and their usefulness in wavelength measurement. He also explored the ramifications of periodic errors in the placement of grating lines. This type of error later bedeviled grating makers, but Fraunhofer found the possibility of periodic error intriguing, because groove spacing that was *"unequal in a regular manner"* might reveal spectral lines not manifest in spectra from relatively perfect gratings. Further, he considered effects that might result from control of the exact shape of the diamond-cut grooves. (Much later, in the twentieth century, this control of groove shape, or "blazing," greatly enhanced stellar spectroscopy.)[11]

Although Fraunhofer's diffraction studies long remained obscure, his other optical work, including the production of lenses and the development of clock drives for telescopes, was well recognized. His skill, together with available resources and favorable political conditions, "gave rise to an optical technology which was to become the envy of the world." At the secularized monastery at Benediktbeuern, the "Benedictine monks had a thousand-year tradition of glass-making." They were also familiar with optics and lens production.[12] With the

---

[10]Ibid., 58 note.

[11]On these insights, see ibid., 56-57, and Ames's "Preface," v, both in *The Wave Theory*. Emphasis in original.

[12]Myles W. Jackson, "Artisanal Knowledge and Experimental Natural Philosophers: The British Response to Joseph von Fraunhofer and the Bavarian Usurpation of Their Optical Empire," *Studies in History and Philosophy of Science*, 25 (1994): 551, 554.

benefit of this expertise, and the expertise of others, and not on account of his diffraction work, Fraunhofer became an eminent figure. But his achievements in diffraction were historically isolated, because he did not train physicists who might have carried on this work, and he did not publish details of his ruling engine. His study of his own tools suggests what one author has called "the reflexive nature of experiment."[13] In spectrum optics, sociologically if not conceptually, Fraunhofer stood at the end of a tradition, rather than the beginning of one.

In the nineteenth century, various machines existed for accurately inscribing multiple lines on hard surfaces. One type marked units of length on measuring scales or degrees in a circle, on instruments for observation and measurement; these "dividing engines" had existed since the eighteenth century.[14] Ruling was also associated with minting currency, and with watch- and clock-making. Other devices, for drawing parallel lines, made ornamental metal surfaces that would show bright, shifting colors by diffraction.[15] Microscopists, too, needed finely ruled parallel lines, as linear scales for the measurement of extremely small objects; dividing engines produced these scales, as well. However, the application of dividing engines to spectroscopic ends waited, after Fraunhofer, until the 1860s, when the Pomeranian instrument-maker Friedrich Nobert created diffraction gratings for Anders Ångström, in addition to the microscopic scales that he made for various users.[16] (Ångström's name was later universally adopted for the standard unit of wavelength measurement, 10^{-10} meters.) The Swedish physicist then used Nobert's instruments to observe the solar spectrum and to create by hand an engraved map exhibiting about 1,000 absorption lines.

After Nobert and Ångström, Americans took an increasing interest in the creation of dividing engines, or ruling engines, as they were eventually called. Lewis Rutherfurd, a wealthy New York amateur astronomer, started to develop a ruling engine in the 1860s. Powered by the flow of water, the engine used a diamond stylus to inscribe lines, and Rutherfurd began to distribute the resulting gratings in the 1870s. The devices, generally no more than two inches across, and ruled with as many as 17,296 lines to the inch, received acclaim from recipients,

---

[13]Matthias Dörries, "Balances, Spectroscopes, and the Reflexive Nature of Experiment," *Studies in History and Philosophy of Science*, 25 (1994): 2.

[14]Allan Chapman, *Dividing the Circle: The Development of Critical Angular Measurement in Astronomy, 1500-1850* (New York: Ellis Horwood, 1990), Chap. 8. Richard John Sorrenson, "Scientific Instrument Makers at the Royal Society of London, 1720-1780" (Ph.D. diss., Princeton University, 1993), 216-224.

[15]Chris Evans, *Precision Engineering: An Evolutionary View* (Bedford, UK: Cranfield Press, 1989), 91-92.

[16]Ibid., 95-96.

some of whom could observe more of the spectrum than ever before. As historian Deborah Jean Warner recounts, Rutherfurd "freely distributed at least fifty gratings to scientists around the world and in so doing made diffraction spectroscopy both possible and popular." (In the 1870s, prisms remained the mainstay of spectroscopy.) Rutherfurd also used the devices himself to photograph the solar spectrum.[17]

A Harvard astronomer, William Rogers, took up the challenge of constructing ruling engines after Rutherfurd, but the principal result of his labors was to provide guidance to Rowland, who began to dominate the art in the 1880s, and whose department of physics maintained its lead into the 1940s. Rogers did not produce instruments in significant numbers, but Rowland examined a Rogers engine, and adopted some of Rogers's techniques, including techniques for the production and use of feed screws. Rowland also obtained tools from Rogers. After Rowland entered the field, Rogers largely relinquished it, and turned to the production of very accurate measuring bars, with which he was able to study grating accuracy.[18] (In fact, most of the engines Rogers created were dual-purpose, and could produce either measuring scales or diffraction gratings.)[19] Evidently there was no rancor over the Baltimore physicist's entry into the ruling art. Rogers wrote to him in 1885, "It is a real pleasure to measure your gratings. It is surprising how nearly you keep your temperature constant." Precise measurement at Harvard required great thermal and mechanical stability. Rogers told of a new measuring room that fostered his metrology: "It is in the underground passage of the Observatory and is surrounded by 1000 tons of granite." He also sold measuring bars to the Hopkins department. Concerning one, he wrote, "I have tried to make it nearly perfect but as usual I have not quite succeeded, but the errors are very well determined and they are pretty small." Thus ruling engine operation was closely allied with precise measures of length.[20]

The process of drawing many near-perfect lines would have been familiar, in principle, to an alumnus of the Rensselaer Polytechnic Institute. The curriculum adopted in 1854 included courses in drawing (under the heading of "graphics") in every semester of study: elementary, topographical, geometrical, architectural, and machine

---

[17]Warner, "Lewis M. Rutherfurd," 192-193, 206, 208, 210, 213. Rowland, "Preliminary Notice of the Results Accomplished in the Manufacture and Theory of Gratings for Optical Purposes" (1882), in *PPR*, 487.

[18]Chris J. Evans and Warner, "Precision Engineering and Experimental Physics: William A. Rogers, the First Academic Mechanician in the U.S.," in *The Michelson Era in American Science, 1870-1930*, ed. Stanley Goldberg and Rogers H. Stuewer (New York: American Institute of Physics, 1988), 4-6.

[19]Evans, *Precision Engineering*, 81.

[20]William A. Rogers to Rowland, 8 February 1885, 27 March 1884, HAR.

drawing. There was also extensive study in geodesy and surveying, including farm surveying, hydrographical surveying, railroad surveying, and mine surveying,[21] all of which afforded Rensselaer graduates extensive experience with lines both ideal and real. Lines ordered the natural and the artificial environments of the engineer, and lines captured the outlines of both nature and machine. Thus Rowland simply automated an operation he had performed countless times himself: the drawing of straight lines.

Of course, gratings required lines so fine, so numerous, so nearly parallel, and so accurately spaced, that production by human hands was out of the question. Automation of production was essential. Thus other modes of mechanical engineering came into play, benefitting from the experiences of both Rutherfurd and Rogers. Rowland improved over Rutherfurd's best results more than twofold, attaining grating dimensions of four inches by more than five inches, with densities as great as 43,000 lines per inch.[22] The magnitude of the dispersion (or magnification) of any given order of the spectrum depended upon the density of lines, and so increased dramatically in the new instruments. The defining, or resolving, power of a grating depended upon the total area ruled and the total number of lines, and indicated the power to make close pairs of spectral lines separate and distinct. Thus this parameter, too, increased dramatically. It was defined as "the ratio of the wavelength to the distance apart of the two spectral lines which can be just seen separate in the instrument."[23] Increases in defining power from more numerous inscribed lines brought increases in the total number of spectral lines observed in the solar spectrum. Coarse spectral lines were split into two or more finer spectral lines. Rowland's instruments thus advanced physical analysis, in the sense of breaking a whole into constituent parts.

The critical problem in the production of high-resolution grating spectroscopes was securing uniformity of spacing between all lines or grooves ruled on the reflecting surface. In the ruling engine, a feed screw ("the simplest instrument for converting a uniform motion of rotation into a uniform motion of translation")[24] caused and regulated the sideways motion of the speculum metal plate, between successive passes of the diamond ruling tip along the plate to scribe lines. The feed screw governed the success or failure of attempts to make all lines parallel and evenly spaced, and for Rowland it had to be perfect.

---

[21]Palmer C. Ricketts, *History of the Rensselaer Polytechnic Institute, 1824-1914* (New York: John Wiley and Sons, 1914), 196-198.

[22]"Preliminary Notice," 487-488.

[23]"Diffraction Gratings" (1902), in *PPR*, 587-588.

[24]"Screw" [undated], in *PPR*, 506.

Making a perfect screw was, he said, a task that "mechanics of all countries have sought to do for over a hundred years and have failed." But Rowland envisioned a way of making an optimal screw, and then designed the remainder of his ruling engine.[25] Innovation in the machine shop was essential to advancements in the analysis of light, and in this sense Rowland's nineteenth-century familiarity with machines yielded fruits important to the twentieth century, and to physicists who might be less conversant with shop technology.

To make the optimal screw, Rowland resorted to the simple method of grinding.[26] With the use of a grinding material such as emery and oil, a very long nut applied to a screw could "grind out all errors of run, drunkenness, crookedness, and irregularity of size." (A screw was "drunk" if a short nut displayed a "wabbling motion" when the screw was turned.) However, the grinding process for a screw of soft Bessemer steel, preferably with a Bessemer nut, required two weeks, during which the nut had to be reversed every ten minutes. Perfection also required that the screw be surrounded by water held at a temperature constant to within 1∞ C. Even the weight of the nut had to be offset.[27] In 1882, Rowland reported that he had reduced errors to below one one-hundred-thousandth of an inch.[28] (Although the screw possessed a very accurate form, the threads were not extremely fine. All three of the ruling engines that Rowland eventually built possessed screws with about twenty threads to the inch.)[29]

Perfection also required a patient human agent, much less visible than Rowland himself, to attend to the tedious grinding process. It was Theodore Schneider "who for twenty-five years was Professor Rowland's mechanician and assistant. . . . It was he who made the screws and most of the working parts of the [ruling] machines."[30] When Rowland taught at Rensselaer, Schneider worked for him through a local firm in Troy, New York. Rowland found Schneider so indispensable that, when he started a new department in Baltimore in 1876, he hired the machinist as an in-house instrument-maker.[31]

The overall ruling engine was not a simple machine like the screw, and therefore design had to compensate for the inevitability of error and imperfection. The removal of all irregularity of motion called for recognizing possible sources of error, and compensating for them in the

---

[25]"Preliminary Notice," 487-488.
[26]"Screw," 506-507.
[27]Ibid., 507-508.
[28]"Preliminary Notice," 488.
[29][Committee], "A Description of the Dividing Engines Designed by Professor Rowland," in *Physical Papers*, 691.
[30]Ibid., 692.
[31]Theodore Schneider to Rowland, 1 September 1876, 21 October 1876, HAR.

overall configuration of the engine: "The secret of success is so to design the nut and its connections as to eliminate all adjustments of the screw and indeed all imperfect workmanship." Rowland also had to provide means of compensating for inevitable wear of the engine.[32]

After the achievement of a practically perfect screw, the next most critical component was the diamond point that inscribed the fine parallel lines. No two points ever had exactly the same shape, but a given diamond usually lasted long enough to rule an entire grating, so that the grooves in one instrument would possess uniform shape. Acquisition of suitable diamond points required both persistence and luck. Rowland's student Joseph Ames reported that "a year may easily be spent in search of a suitable diamond-point." Unsuitable diamonds were most common: "Most points make more than one 'furrow' at a time, thus giving a great deal of diffused light." The behavior of diamonds also made ruling on glass largely impracticable, for glass brought the added difficulty "of the diamond-point continually breaking down."[33]

The overall power for the ruling engine came from a water-motor, via a large pulley. The rotation of this pulley and its shaft in turn caused the several motions of the ruling engine. (See Figure 2.) A crank on the main shaft moved a connecting rod attached to the housing of the diamond point. As the diamond moved, it inscribed a line on a speculum metal plate. After the completion of one stroke (the length could be adjusted, up to 4 1/4 inches),[34] a cam caused the diamond to lift from the metal surface before it returned to its starting position. While the diamond was raised, other cams on the main shaft moved pawl-levers which caused a slight rotation of a very large gear, or ratchet-head, concentric with the feed screw and attached to it, thus rotating the screw very slightly. A nut on the critical feed screw imparted a slight motion to the carriage that held the speculum metal plate, so that the next stroke of the diamond would rule a line a few ten-thousandths of an inch separated from the previous line.[35]

Once in operation, the ruling engine could not be stopped before the grating was complete, because the process of starting and stopping would inevitably cause departure from uniform spacing. Thus, for example, once the engine started to rule a grating of six inches in width, having a density of 20,000 lines to the inch, Rowland or his assistants were committed to running the engine for five continuous days and nights.[36] (Although Rowland never made the comparison,

---

[32]"Screw," 509-510.
[33]Ames, "The Concave Grating in Theory and Practice," *Philosophical Magazine* 27 (1889): 383-384.
[34]"Preliminary Notice," 490.
[35]"A Description of the Dividing Engines," 691-693.
[36]Ames, "The Concave Grating," 383.

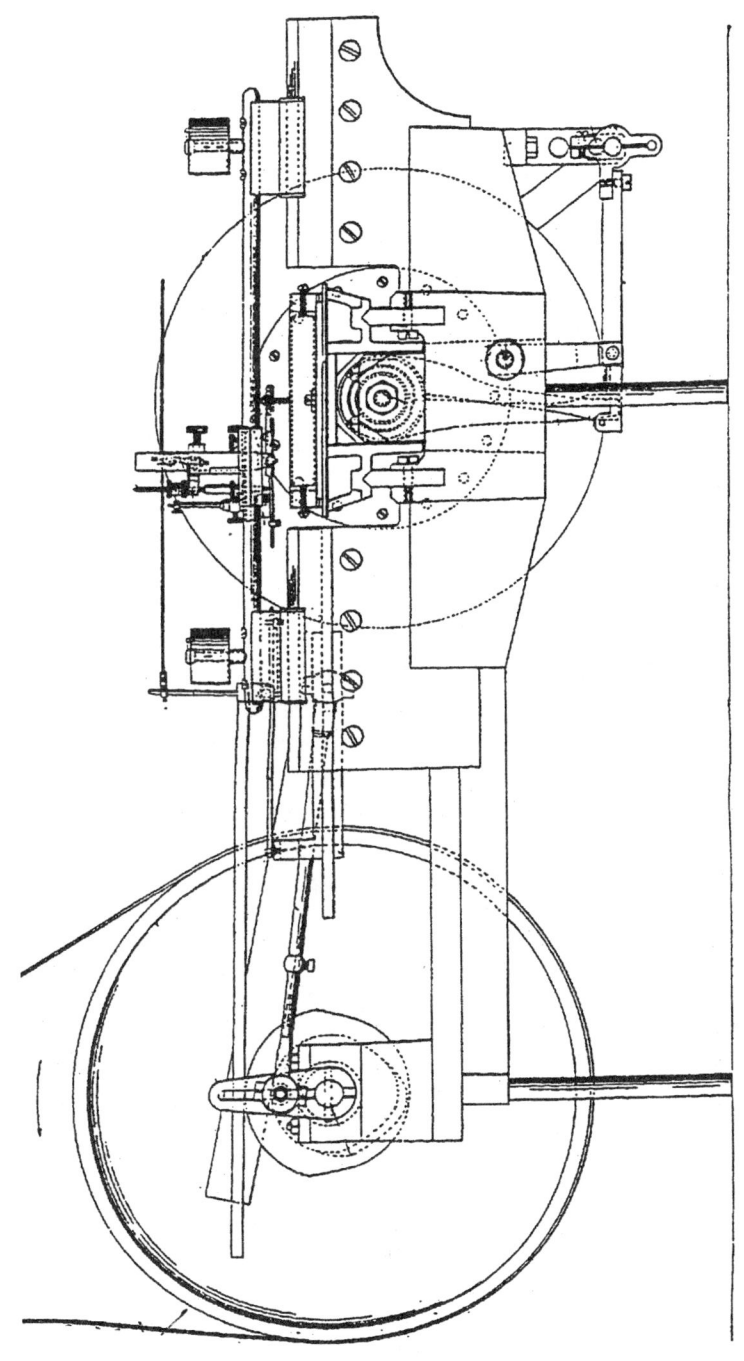

FIGURE 2. *Side view of Rowland II ruling engine. It received power via the belt and large pulley at left. The diamond point is shown resting on a grating in production at right. From PPR.*

his diamond ruled lines much as a stylus inscribed lines on Thomas Edison's recording phonograph of 1878, with important exceptions: that vibrations were excluded from Rowland's engine while they were essential to Edison's recording machine, and that the ruling engines had to run for much longer times at each instance.) Engine operation, like the production of a perfect screw, required the attentions of an assistant. Here again, Schneider was Rowland's right-hand man: "It was he who superintended the ruling of every grating that has left the Physical Laboratory of the Johns Hopkins University for use elsewhere in the world."[37]

As well as continuous operation, the ruling engine required extreme thermal and mechanical stability. A reporter for the *Baltimore American* found the ruling engine running inside a box that protected it from changes in temperature. He relayed his experience that "the courteous attendant, previous to removing the door to the box, informs you that he can let you look in but a moment or two, as the presence of the human body for a longer time would be ruinous to the accuracy of the work."[38] Rowland kept the engine in an underground vault to eliminate vibrations from traffic in the street and to help keep the temperature constant.[39]

Partly because of the great delicacy of the ruling process, and partly because the ruling engines were adjustable, no two gratings were ever exactly alike. In this sense, the creation of gratings resembled American mass production less than it resembled the older system known as "making," in which each product is slightly different from all others.[40] The line spacing, the length of lines ruled, and the width of the ruled area could all be controlled, within limits, but the exact shape of the diamond point was partly a matter of luck.

Once completed and placed in use, every concave grating simultaneously produced multiple, overlapping spectra of a given source, of different "orders," corresponding to optical path lengths differing by one wavelength, two wavelengths, and so on. High-order spectra possessed greater dispersion than low-order spectra. But the brightness of any given order depended on the particular grating producing it, with great variation from one grating to the next, depending

---

[37]"A Description of the Dividing Engines," 692.

[38]*Baltimore American*, 18 November 1883, in Scrapbook, Manuscript Collection 1045, American Philosophical Society, Philadelphia.

[39]John A. Brashear, *John A. Brashear: The Autobiography of a Man Who Loved the Stars*, ed. W. Lucien Scaife (New York: American Society of Mechanical Engineers, 1924), 74.

[40]On the distinction between making and manufacturing, see Paul Uselding, "Measuring Techniques and Manufacturing Practice," in *Yankee Enterprise: The Rise of the American System of Manufactures*, eds. Otto Mayr and Robert C. Post (Washington, DC: Smithsonian Institution Press, 1981), 109-110.

upon the exact shape of the grooves, which depended upon the exact shape of the diamond stylus that made them. Rowland once observed, "No two gratings are ever alike in this respect, but exhibit an infinite variety of distributions of brightness."[41]

This variation was not an absolute liability, if the performance of each instrument was known, because the order of the spectrum desired in a given series of observations depended upon the region of the spectrum chosen for study. A grating could be selected for its brightness in a specific order at specific wavelengths. Nor did Rowland consider the overlapping of spectra of different orders a problem, because the coincidence provided an opportunity for cross-checking. He noted, "By micrometric measurement of such superimposed spectra we have a most beautiful method of determining the relative wave lengths of the different portions of the spectrum." The investigator could make quantitative comparisons of lines appearing simultaneously in different orders, bringing measurements of different regions into better uniformity. Overlapping spectra even possessed aesthetic appeal he observed, "It is a most beautiful sight to see the lines appear colored on a nearly white ground."[42]

There was one high-precision task in the production of gratings that did not take place at Johns Hopkins, namely the figuring of the speculum metal (copper-tin alloy)[43] plates with very slightly concave surfaces upon which the lines were ruled. The ruled surface never greatly exceeded six inches by four inches, and the radius of spherical curvature for that surface was typically five to twenty feet. Thus the center of a concave region six inches across, with a twenty-foot radius of curvature, would only lie about 0.02 inches, or 0.5 millimeters, lower than the edges. The task of forming this precision surface was given to a Pittsburgh instrument-maker who had for years pursued lens-making as a hobby, taking months or even years to shape each lens on his own time, in addition to his job as a millwright in a Pittsburgh iron works.[44]

John A. Brashear attracted Rowland's attention after he polished a prism for Charles Hastings, who was for a while a colleague of Rowland in the Johns Hopkins physics department and an expert, as Rowland initially was not, in optics. Shaping of the plates for ruling required accuracy of form to within one-fifth of a wavelength of light,

---

[41]"Diffraction Gratings," 588.
[42]"Preliminary Notice," 489.
[43]"A Description of the Dividing Engines," 691.
[44]As well as Brashear's autobiography, *John A. Brashear*, see the derivative work, Harriet A. Gaul and Ruby Eiseman, *John Alfred Brashear: Scientist and Humanitarian, 1840-1920* (Philadelphia: University of Pennsylvania Press, 1940), 39-43.

or one two-hundred-thousandth of an inch. Brashear wrote that, in accepting the initial order, "we had undertaken a task that proved well-nigh capable of flooring us." Brashear rose to the challenge, however, and initially Rowland was satisfied. But when Rowland compared Brashear's plates with test plates from the German firm Steinheil, he found that the former were misshapen. After Brashear made the trip to Baltimore to discuss the problem, however, the Steinheil test plates, rather than Brashear's, were found to be in error, and Rowland thenceforth accepted Brashear's work without question.[45]

Brashear became a full-time instrument-maker at about the same time that he began figuring the plates. But because he had little patience for cost accounting, a Pittsburgh philanthropist, William Thaw, made his work possible by paying him a salary, and simultaneously urging Brashear to obtain payment from everyone who acquired the various instruments he produced.[46] The work for Rowland, Brashear remembered, "opened a new field for us, and not among amateurs, but among the most advanced scientific men of the day."[47] For William Rogers, Brashear made half-meter bars, straight to within one fifty-thousandth of an inch. Rogers then ruled these bars to create accurate measuring scales.[48] For A.A. Michelson, Brashear made a rotating-mirror apparatus, able to withstand 50,000 revolutions per minute, for measuring the speed of light,[49] and optical surfaces for Michelson's first interferometer. Brashear was also involved in creating equipment by which to measure the standard meter bar in Paris in terms of wavelengths of light.[50] For Samuel Langley, Brashear constructed rock salt prisms, needed for studies of infrared light in Langley's new bolometer, and eventually many parts for Langley's experiments with heavier-than-air flying machines.[51] Spectroscopes and large lenses for many observatories also came from the Pittsburgh maker.[52] Brashear was

---

[45]Brashear, 74-75, 77-78. Later, in marketing grating spectroscopes after the turn of the century, the German firm C.A. Steinheil Sohne incorporated either Rowland plane gratings or copies of gratings, or included no gratings at all, evident testimony to the primacy of Rowland gratings. *Price-List of Astronomical and Physical Instruments* (Munich: C.A. Steinheil Sohne, 1907), 67-68.

[46]Gaul and Eiseman, 81.

[47]Brashear, 76.

[48]Ibid., 90.

[49]Gaul and Eiseman, 85. Later, in the 1920s, Michelson relied on the Sperry Gyroscope Company for apparatus used in a redetermination of the speed of light at Mount Wilson Observatory in California. Thomas Parke Hughes, *Science and the Instrument-Maker: Michelson, Sperry, and the Speed of Light* (Washington, DC: Smithsonian Institution Press, 1976), 1, 4.

[50]Brashear, 245.

[51]Brashear, 80-81, 130-131, 243-244. Rowland, too, considered the possibility of flying machines. While still a student, he concluded that "a little more than half a horse power" would be required to permit one person to fly. He hoped to refine the estimate. "Volume 1," 1868 notebook, 77, Box 20, HAR.

[52]Brashear, 245-247.

content in these matters simply to be useful. As two biographers stated, "He was the toolmaker. He let others have the credit of being scientists."[53] (He actually received much acclaim as an instrument-maker during trips to Europe.)

With the material assistance of Schneider and Brashear, Rowland was very soon able to report successes in production, making both concave and plane gratings. He initially stated, "Every grating made by the machine is a good one . . ., but some are better than others." This was a great improvement over Rutherfurd, whose machine "only made one in every four good, and only one in a long time which might be called first-class." From the beginning, the Hopkins operation produced both concave and flat gratings, in sizes ranging from one inch square to, initially, 4 by 5 3/4 inches, although the highest capacity of the engine was 4 1/4 by 6 1/4 inches.[54]

The Rensselaer alumnus reported line densities ranging from 14,438 to more than 43,000 per inch, and found his ruled lines to be straight to within one hundred-thousandth of an inch, the same as the accuracy of his feed screw. Different gratings worked best in different orders of the spectrum, and observations were reported up to the tenth or twelfth order. Rowland also reported the detection of many spectral lines never distinctly seen before.[55] With assistance, he photographed and re-photographed the solar spectrum, and set out to measure all solar absorption lines—"an appalling task which was to consume the best part of his energies for the next fifteen years."[56] Still, with photography he eliminated the task of drawing by hand.

With the choice of gratings over prisms, Rowland also eliminated many difficult calculations. One simple multiplication converted line positions into wavelengths, and eventually he made a "measuring engine" to eliminate even that one step. The engine, he wrote, "measures *wave-lengths direct*, so that no multiplication is necessary, but only a slight correction to get figures correct to 1/100 of a division of angstrom."[57]

The importance of instrument output from Baltimore became apparent even before Rowland publicized his solar studies. Samuel Langley, director of Allegheny Observatory in Pennsylvania, performed one of the most dramatic first uses of the concave grating. His observatory received income for providing a time service to railroads and cities, and Langley drew on this revenue to aid his studies of the

---

[53]Gaul and Eiseman, 82.
[54]"Preliminary Notice," 490-491.
[55]Ibid.
[56]Crew, *The Rise of Modern Physics*, 313.
[57]"Report of Progress in Spectrum Work" (1891), in *PPR*, 521. Emphasis in original.

sun.[58] He used Rowland's device in measuring the distribution of energy across the visible solar spectrum, and well into the infrared region.

For the task, Langley invented the bolometer, which utilized both a Rowland grating and a prism of either glass or rock salt, the latter made by Brashear. The two components together isolated lines of the solar spectrum. Each minuscule segment of the spectrum then fell on, and slightly heated, a strip of platinum less than 1/25,000 inch in thickness and 1/125 inch in width. The temperature in turn affected the amount of electric current running through the platinum strip. Langley reported that a galvanometer in circuit with the strip could detect temperature changes of much less than 1/10,000 degree Fahrenheit. He reported both his methods and his findings to a meeting in Southampton, England:

> Since it is one and the same solar energy, whose manifestations we call 'light' or 'heat,' according to the medium which interprets them, what is 'light' to the eye is 'heat' to the bolometer, and what is seen as a dark line by the eye is felt as a cold line by the sentient instrument.[59]

From measurements with the galvanometer across the spectrum, he constructed a curve showing the distribution of energy at all detectable wavelengths. (See Figure 3.) But the derivation was not simple. Langley wrote, "The whole work, it will be seen, is necessarily very slow; it is in fact a long groping in the dark."[60]

He paid special attention to the infrared region of the spectrum, which he believed had been "singularly underestimated" because of "the deceptively small space into which it appears to be compressed by the distortion of the prism." The compression made wavelength determination more difficult: "The most tedious part of the whole process, has been the determination of wave-lengths." Knowledge of infrared wavelengths was, until his own work, mostly indirect, Langley said, inferred as it was from formulas, "all which known to me appear to be here found erroneous by the test of direct experiment." The "direct" observation was made possible by the new Hopkins instrument, for, he said, "I have been greatly aided in this part of the work by the remarkable concave gratings lately constructed by Prof. Rowland of Baltimore."[61]

The investigations also led into physics, as Langley attempted to evaluate the re-emission of solar energy from planetary surfaces. To

---

[58]Charles D. Walcott, "Samuel Pierpont Langley, 1834-1906," *BMNAS*, 7 (1913): 248.
[59]Langley, "Sunlight and Skylight at High Altitudes," 587.
[60]Ibid.
[61]Ibid.

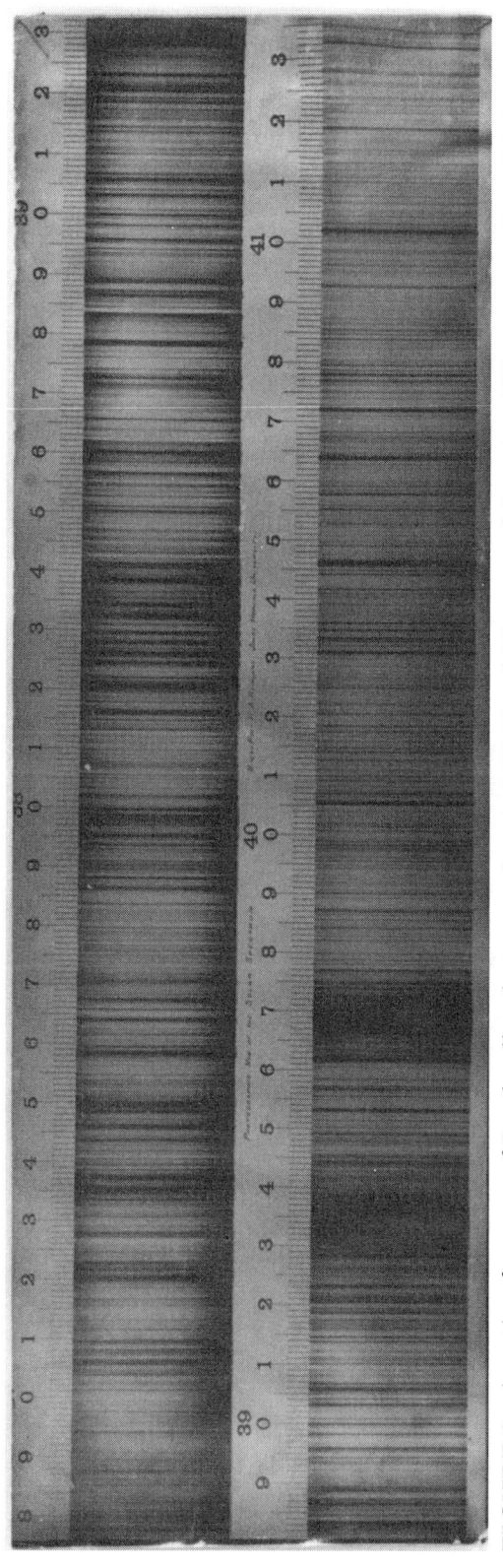

FIGURE 3. *A section from one of Rowland's solar maps. Courtesy National Museum of American History, negative #64581.*

that end, he experimented with thermal emission from copper, coated with lampblack, in the laboratory, mostly in the infrared, and he compared the laboratory curves he derived with his solar curves. As Thomas Kuhn described these laboratory studies, "Langley's experiments [were] the mere beginning of the work on which the development and evaluation of quantitative black-body laws would depend."[62] As for the future of solar studies, Langley believed that photography, probably by others, would ultimately reveal more about the sun than his galvanometer needle, and placed his confidence in "the camera, an instrument, which where it can be used at all, is far more sensitive than the bolometer."[63]

In 1882, the year of Langley's talk at Southampton, Rowland visited Europe himself, and the physicist and his invention were greeted as an American wonder—"like the Yosemite, Niagara, Pullman parlor car."[64] He spoke before the Physical Society of London, and announced his dramatic new means of dividing the spectrum and establishing wavelengths: "Nothing can exceed the beauty and simplicity of the concave grating" when suitably mounted. Photographing long sections of the normal spectrum became a simple matter. He said, "Thus the work of days with any other apparatus becomes the work of hours with this. Furthermore, each plate is to scale, an inch on any one of the strips representing *exactly* so much wave-length."[65] *Nature*, the journal founded in England by J. Norman Lockyer in 1869, reported: "Lines are divided by this method which have never been divided before; and the work of photographing takes a mere fraction of the time formerly required." The journal also mentioned the engine that made the gratings: "An account of this machine will be published shortly."[66] However, the engineering physicist must have decided subsequently that there was little to gain from disclosure, for mechanical diagrams of the engine did not actually appear until 1902.

Rowland believed that he had very nearly attained practical perfection, as measured by the sharpness of the spectral lines produced by his gratings. In his analysis, the actual width of any given spectral line depended upon the number of linear grooves ruled on the grating, upon the width of the aperture admitting light onto the grating, and upon "the *true physical width of the line*," something independent of the spectroscopist's, or the instrument-maker's agency. Very soon

---

[62]Kuhn, *Black-Body Theory*, 7-9.

[63]Langley, "Sunlight and Skylight," 587.

[64]John Trowbridge to Daniel Coit Gilman, 30 November 1882, in *Science in Nineteenth-Century America: A Documentary History*, ed. Nathan Reingold (London: Macmillan, 1966), 273.

[65]"On Concave Gratings for Optical Purposes" (1883), in *PPR*, 499-500. This article, written *in extenso*, was based on the 1882 London address, which was essentially an abstract. Ibid., 492.

[66]"Physical Society" [report of meeting], *Nature*, 27 (1882): 95.

after he began to distribute gratings, Rowland believed that his instruments approached an ideal, giving resolution for the solar spectrum which could not be exceeded "by even a spectroscope of infinite power." However, "we should not expect any *definite* limit, but a gradual falling off as we increase our power." Probably the smallest detectable separation of two lines was 1/150,000 of the wavelength. He opined that "we can never hope to see very many more lines in the [solar] spectrum than can be seen at present, either by means of prisms or gratings."[67] (Rowland published his last compendium, of some 20,000 absorption lines of the solar spectrum, expressed to 0.001 angstrom units, in the late 1890s. By the 1960s, about 4,000 additional absorption lines were known.)[68] In effect, Rowland argued that he had created the *ne plus ultra* of spectroscopic instruments.

The fortunes of the physicist rose with those of his invention. Harvard physicist John Trowbridge was present in London for the address to the Physical Society, and reported to Gilman, "The enthusiasm was tremendous. Johns Hopkins University was lauded to the skies. Rowland has already made himself immortal. I felt very patriotic as I listened to him and saw the astonishment of the Englishmen. Our visit has been profitable in various ways."[69] The simplicity of measurement which accompanied use of the instrument was especially impressive. It was a labor-saving device:

> When he said that he could do as much in an hour as had hitherto been accomplished in three years, there was a sigh of astonishment and then cries of "Hear! Hear!" Professor [James] Dewar arose and said "We have heard from Professor Rowland that he can do as much in an hour as has been done hitherto in three years. I struggle with a very mixed feeling of elation and depression. Elation for the wonderful gain to science and depression for myself, for I have been at work for three years in mapping the ultra violet."[70]

Experiment became easier, and therefore experimental results could be more abundant. This achievement impelled Rowland's "great triumphant march" in Europe. Trowbridge himself could not remain disinterested; he told Gilman, "I want you to use your influence with him to give me one of his best gratings; I am influenced both by selfishness and by a desire to secure the Rumford Medal for him." Trowbridge said he could generate more support for the award with a Rowland grating in his possession.[71] (He later got his wish;

---

[67]"On Concave Gratings," 502-504. Emphasis in original.
[68]Hearnshaw, *The Analysis of Starlight*, 422.
[69]John Trowbridge to Gilman, 12 November 1882, in *Science in Nineteenth-Century America*, 272.
[70]Trowbridge to Gilman, 30 November 1882, 273.
[71]Ibid., 272, 274.

Johns Hopkins presented a grating to him and to Harvard College in February, 1884.)[72]

The Baltimore physicist also demonstrated his creation in Paris to Sir William Thomson, French physicist E.E. Mascart, Italian physicist F. Rossetti, and German physicists F.W.G. Kohlrausch and G.H. Wiedemann. In Trowbridge's account, "They were dumbfounded. Mascart kept repeating with bated breath 'Il faut qu'on commenserat [sic] encore' or words to that effect."[73] The German physicists "spread their palms, looked as if they wished they had ventral fins and tails to express their sentiments. . . . We left the French capitol [sic] with the feeling that there was little to be learned there in the way of physical science." Thomson, Mascart, and the others were to be "heralds to proclaim the preeminence of American diffraction gratings."[74]

The instruments quickly began to travel independently of their maker. James Keeler, who completed undergraduate studies in physics and mathematics at Johns Hopkins in 1881, brought a Baltimore grating with him to Germany in 1883, after working for two years with Langley at Allegheny Observatory. In Heidelberg, Keeler attended the lectures of Georg Quincke on light. During one lecture, Quincke showed them a grating made by Friedrich Nobert, "saying it was one of the largest and most perfect gratings in existence." Keeler spoke up to say that he had a grating three times as large, but Quincke doubted it was usable. After Keeler retrieved the device from his room and displayed it, Quincke "was astonished with its apparent beauty. His skepticism turned to unstinted praise."[75]

By the end of 1883, recipients of Rowland gratings included the prominent English spectroscopist J. Norman Lockyer, German physicist Hermann von Helmholtz, E.E. Mascart, John Trowbridge, Sir William Abney, A.G. Stoletow of the University of Moscow, and Otto Struve at Pulkowa Observatory (not to be confused with his grandson, Otto Struve, who emigrated to America). Still, like Rittenhouse, Fraunhofer, and Rutherfurd, Rowland foresaw little or no monetary profit for himself from diffraction instruments, despite his promotional efforts, which concerned renown and authority more than money. Rowland once informed Gilman about the instruments, "There are a great many on hand at present and I would suggest that they be given away." Although he gave a few gratings away, he also made an arrangement under which Brashear "agreed to polish the plates

---

[72]Record of grating, Inventory number 2031, Harvard University Collection of Historical Scientific Instruments, Cambridge, MA.

[73]Trowbridge to Gilman, 12 November 1882, 271-272.

[74]Trowbridge to Gilman, 30 November 1882, 273.

[75]Brashear, 78. Donald E. Osterbrock, *James E. Keeler: Pioneer American Astrophysicist* (Cambridge: Cambridge University Press, 1984), 31.

hereafter and receive half the profits, 10 per ct going to Schneider for attending the machine at night."[76]

When an arrangement was formalized at the end of 1884, Brashear became Rowland's official marketing agent, agreeing to send Johns Hopkins one-half of gross receipts, and to cover his expenses out of the half he kept. He also agreed to "attend to all the correspondence and business connected with the sale of the gratings."[77] Rowland had originally hoped that Schneider would market the instruments, and give the university part of the money, which would then support the mapping of the solar spectrum,[78] but apparently Brashear was much better equipped for the task.

Rowland once had to defend Schneider's role to President Gilman. He wrote, "It must be remembered that the machine is *unique* and so gratings cannot be obtained except here. Besides, it will not look badly to see in every important paper on the spectrum, the name of this University."[79] Five years later, Rowland explained the necessity of Schneider's work again:

> Mr. Schneider is an absolute necessity to the Physical Laboratory. . . .
> The engine cannot be run without a person near it all the time to oil
> it & keep it in repair, to put the belts on & off the dynamos & to see
> that no accident happens. The running of the engine is necessary for
> the spectrum work to which the whole top story of the laboratory is
> devoted. . . . If Schneider is to go, all spectrum work must cease.[80]

The implications were global: "This is the only place in the world where they can be made well & every spectroscopist in the world is dependent on them. It would be a considerable blow to science if this was interfered with. There are people all over the world waiting for them." Rowland simultaneously assured Gilman, "Schneider now gives only his odd time to the gratings," and operation of the engine was not his principal duty. His principal function was to provide apparatus for original experiments of various kinds. There was also hope that revenue from the diffraction instruments could cover general shop expenses. No longer talking of giving instruments away, Rowland now proposed raising prices. But even if the operation ran a slight deficit, he argued, Johns Hopkins should be willing to cover the difference.[81]

Between 1882 and 1887, expenses related to grating production, including commissions paid to Schneider but not his base salary,

---

[76]Rowland to Gilman, 1 January 1884, HAR.
[77]Executive Committee agreement, 20 December 1884, in Correspondence - Rowland, DCG.
[78]Rowland to Gilman, 25 September 1882, DCG.
[79]Rowland to Gilman, 1 May 1883, DCG.
[80]Rowland to Gilman, 16 June 1888, HAR.
[81]Ibid.

totaled $1,336, compared with receipts of $2,392. After Rowland's plea to Gilman, the amount paid to the machinist for grating production increased, and between 1888 and 1901, the operation generally ran a deficit, with receipts running about eighteen percent below expenses. (Annual receipts averaged about $850.)[82] Evidently Schneider was largely uncomplaining; he even minded Rowland's house when the physicist was elsewhere. But Schneider's health did not survive all challenges. He reported in 1897, "As to the [ruling] vault, I have worked there when the weather was cool—the doctor would not allow me to work there when the weather was warm." When Rowland himself was ill later that year, Schneider advised, "Everybody is writing for gratings. I had to give up the new diamond and go back to the old one and now have a point that I think will do." The (presumably uniform) heat and humidity of the ruling vault in the summertime was apparently the most difficult condition for Schneider.[83] (The ruling vault was actually the site of Rowland's eventual interment. The ruling engines were so important to him that, after his death, his ashes were placed in a niche in the vault wall.)[84]

While Schneider bore much of the burden of the ruling operation, John Brashear received much of the acclaim. The device that gave Rowland such a welcome in Europe also opened doors for Brashear, who brought a prime six-inch specimen with him in 1888. During his travels, to avoid damage by curious customs inspectors, he wrapped the instrument in chamois and kept it in his pocket. As he recalled, "It weighed at least ten pounds, and I am sure my overcoat on that side was two inches longer than on the other before I got through with my journey."[85] Jules Janssen at Meudon Observatory had corresponded with Brashear about gratings, and now provided a tour of the observatory, showing him "the spot where he started off in a balloon loaded with his observing outfit, and sailed over the German army to Algiers, for the purpose of observing the total eclipse of the sun."[86] Another pleased customer was the English astronomer Piazzi Smyth, who showed Brashear his private laboratory.[87]

Victor Schumann of Leipzig, still another user, shared photographs of the ultraviolet spectrum as far as 1850 angstroms.[88]

---

[82]"Gratings a/c," in Correspondence - Rowland, DCG; "Receipts and Expenses, a/c Gratings from November 1888 to January 31, 1901," Box 15, HAR.

[83]Schneider to Rowland, 9 August 1897; Schneider to Mrs. H.A. Rowland, 17 August 1897; Schneider to Rowland, 18 July 1900, HAR.

[84]*John A. Brashear*, 79.

[85]Brashear, 98.

[86]Ibid., 107-108.

[87]Ibid., 125.

[88]Ibid., 111.

Brashear was already aware of the amateur's enthusiasm for the new instrument. The prior year Brashear had reported to Rowland: "Previously he worked with a battery of 18 quartz prisms. . . . [W]hen your grating was received he was the most enthusiastic German I ever heard from. Got up out of a sick bed to see the spectra & said 'his [sic] eyes were *glued* to it for two hours.'"[89] During a later trip to Britain, in 1892, at a dinner of the Royal Astronomical Society, Brashear "could not help but have an occasional thought that it was only a few years since [he] was a greasy millwright in Pittsburgh."[90] Much of the difference was due to the gratings he marketed (See Figure 4).

While praise for Rowland instruments was nearly unanimous, there were physicists who also felt dismay. As Chris Evans has observed, already by the middle of the nineteenth century "the capabilities of the scientific mechanicians exceeded the ability of scientists to use or comprehend the data provided by the rulings at their disposal."[91] And Rowland gratings only made the quandary worse. For example, Norman Lockyer suddenly faced spectra far more complex than he could have wished. High dispersion disambiguated multiple lines where low dispersion had shown merely single lines, and "this increasing ability to split lines, which he had supposed to be single, upset Lockyer." The complexity undermined his sense of understanding of the origins of spectra. Dewar and a colleague found in the 1880s that some spectral lines, "when examined under a high enough dispersion, proved not to be single at all, but doublets or multiplets."[92] Already in 1882, Trowbridge in London had observed, "Lockyer evidently is much disturbed and was not at the meeting of the Physical Society. He does not impress me as a gentleman. Did not call on us in London."[93]

However, eventually Lockyer came around. In the late 1880s, he acquired a Rowland grating,[94] and his dismay changed to enthusiasm: "Thanks to the generosity of Mr. Rutherfurd and the skill of Professor Rowland, magnificent diffraction-gratings have been spread broadcast among workers in science, and we have therefore easy means of obtaining with inexpensive apparatus a spectrum of the sun." Lockyer envisioned the eventual attainment of "gigantic" solar spectra, extending more than 300 feet, "though certainly not in our time." Lockyer studied limited sections of the spectrum on this scale,

---

[89]Brashear to Rowland, 8 February 1887, HAR.

[90]Brashear, 119-120.

[91]Evans, *Precision Engineering*, 6.

[92]A.J. Meadows, *Science and Controversy: A Biography of Sir Norman Lockyer* (Cambridge, MA: MIT Press, 1972), 154, 155.

[93]Trowbridge to Gilman, 30 November 1882, 274.

[94]Meadows, *Science and Controversy*, 155.

[95]J. Norman Lockyer, *The Chemistry of the Sun* (New York: Macmillan and Co., 1887), viii.

and found that it was "really not beyond what is required for honest, patient work."[95]

But in the 1890s, the new instrument still undercut Lockyer's theoretical views on the causes of spectra. Rowland reported that, at high dispersion, lines which Lockyer thought to be basic were not so; rather, they were "widely broken up and cease[d] to exist." He found that Lockyer "mistakes groups of lines for single lines or even mistakes the character of the line entirely." Against Lockyer's idea that elements come apart in the intense heat of the sun (the "dissociation hypothesis"), Rowland concluded, "There seems to be very little evidence of the breaking up of the elements in the sun as far as my experiments go."[96] Rowland's assistant, Lewis Jewell, also concluded in the 1890s that Lockyer's explanations of spectra were not required for an understanding of the solar spectrum.[97]

Langley had stated in 1882, "The solar spectrum is . . . commonly supposed to have been mapped with completeness."[98] However, his study with the new instrument suggested that much remained to be done. Rowland did not attain the 300-foot spectrum that Lockyer envisioned, but he made available sixty-foot maps, at a scale of about five angstroms to the inch, in three-foot sections. In 1886, his university sold maps that he claimed were "more exact and giv[e] greater detail than any other map now in existence."[99] When the Baltimore physicist converted his maps into tables of wavelengths, he claimed "at least ten times the accuracy of any other determination."[100] Thus the leading word on the composition of sunlight passed from Europe to the United States, where the process of recording the spectrum was simplified and partly automated. And like the gratings, the solar maps traveled, too. More than eighty were distributed, and they reached Germany, England, France, Canada, Scotland, Ireland, Russia, Japan, Spain, Denmark—and numerous institutions in the United States.[101]

Although the gratings and maps arrived in many places, some exemplary uses of the Rowland instrument took place in the Johns Hopkins physics department itself. The most exemplary use was the creation of the maps. In taking spectra of the sun, and also of the elements, Rowland relied on "the faithful and careful work of Mr. L. E. Jewell." As the work continued over a number of years, it was Jewell who made most, by far, of the actual measurements, and who by 1893 became "so expert as to have the probable error of one setting about

---

[96]"Report of Progress in Spectrum Work," 524.
[97]Meadows, 156.
[98]Langley, "Sunlight and Skylight," 586.
[99]"Photograph of the Normal Solar Spectrum" (1886), in Box 35, HAR.
[100]"On the Relative Wave-Length of the Lines of the Solar Spectrum" (1887), in *PPR*, 512.
[101]"Rowland's Photographic Maps (Order List)," Box 15, HAR.

1/1000 division of angstrom, or 1 part in 5,000,000 of the wave-length."[102] However, as Rowland later told his student Henry Crew, anyone who received pay for spectrum research could not also expect to receive academic credit for it.[103] The tables traveled under Rowland's name alone.

To study light, the physicists and staff at Hopkins frequently worked in darkness, even during the day. The room used for comparative solar and terrestrial spectroscopy at Johns Hopkins had blackened fixtures and walls. The window-glass was deep red, and the shades were black. A hood of black cloth prevented unwanted, stray light from entering the narrow aperture into the big spectroscope. Another black hood isolated the camera box from any light other than the spectrum, "as even the darkest room has some light in it." The laboratory was a light-isolation chamber, which permitted only specific forms of light to enter the spectroscope.[104]

A heliostat, a mirror outside on a balcony, controlled the entry of daylight into the darkened laboratory.[105] Any secondary light from the illuminated atmosphere or illuminated objects on the ground was unwanted. In this sense, Rowland's goal was purity and order; where incoming light lacked order, he imposed it, arranging in linear sequence the many colors that made up the yellowish light of the sun. When not studying the celestial, Rowland and his students studied terrestrial matter in artificial purity, one element at a time. The goal was to make the elements, piecewise, shed light where the solar spectrum held dark gaps. The known natural elements, which in Rowland's day were still growing in number, were commanded to illuminate man's ignorance of solar composition.

From the heliostat, sunlight passed through a condensing lens and a totally reflecting prism, both made of quartz, then into the laboratory. An absorbing solution, placed between the lens and prism at the pull of a string, could limit the range of the solar spectrum being admitted. To switch from solar to laboratory observation, the observer could also use a string to cause the prism to deflect sunlight away from the spectroscope. Inside, in a compartment separate from the spectroscope, two artificial light sources were available to illuminate the spectroscope. One was an arc source that received current at 150 volts from one dynamo or alternating current from another dynamo at 700 volts. The second source was a separate spark apparatus that

---

[102]"Report of Progress in Spectrum Work" (1891), in *PPR*, 524; "A New Table of Standard Wave-Lengths" (1893), ibid., 545.
[103]Crew, Diary, 18 November 1886, HC.
[104]Ames, "The Concave Grating," 374, 377.
[105]Ibid., 374-375.

received power from an alternating Siemens dynamo, an induction coil, and three to twelve Leyden jars of one gallon each. A wooden tube with a condensing lens conveyed either of these two kinds of artificial light to the spectroscope.[106]

The principal structural components of the spectroscope, which was built for solidity, were two six- by thirteen-inch wooden beams, each twenty-three feet in length, joined at one end to form a right angle. The precise orientation of one beam could be adjusted by means of screws, to keep the beams exactly perpendicular to each other. Light entered the spectroscope through a slit at the vertex of the right angle, and traveled along one arm to the concave diffraction grating, which diffracted and reflected it, along a diagonal girder, to a photographic plateholder riding on the other arm. (See Figure 5.) Micrometer screws controlled the precise width of the entrance slit, which was generally less than one-thousandth of an inch. Rotation of the slit about a central axis could make it parallel to the rulings on the grating, within 0.5 degrees of alignment.[107]

At the high dispersion that Rowland used, a photographic plate received only a small portion of any spectrum, and Rowland could select the portion by simultaneously moving both grating and plateholder, which were kept a fixed distance apart by the diagonal wrought-iron girder. Each of the two devices rested on its own separate iron carriage with two brass wheels, running on an iron track atop a wooden arm. Essentially, the girder connecting the grating and plateholder formed the moveable hypoteneuse of a right triangle, with sides constituted by the two heavy wooden beams.[108]

Because Rowland adjusted the length of the girder to coincide with the focal distance of the grating, the spectrum remained in focus while the grating and plateholder moved, and swept across the plateholder, until the spectral region to be studied fell upon the photographic plate. The grating focused the overall spectrum on a hypothetical circle whose diameter equaled the radius of curvature of the grating, often 21.5 feet. Of the 67.5-foot circumference, only three feet or less could be photographed at one time, due to the limited size of the photographic plates. The spectroscope had to hold each photographic plate flexed to match the curvature of the focal circle. Wooden buttons in the "camera box" pressed the plate against hard rubber to maintain the necessary shape; a plate thickness of only 1/14 inch allowed the necessary flexibility. Rowland, who treated the plates with an ammonia emulsion, envisioned spectroscopy of the stars eventually, on paper or celluloid film,

---

[106]Ibid., 376.
[107]Ibid., 375-376 and figures.
[108]Ibid., 375.

FIGURE 4. *Diffraction grating, marked: "Plate polished and corrected by J. A. Brashear, Pittsburgh." Courtesy National Museum of American History, negative # 645880.*

which could withstand the more extreme curvature necessary with shorter-focus (hence brighter) concave gratings.[109]

The position of the camera box could be adjusted to precise alignment with the spectrum, which ranged from one-quarter to four

---

[109]Ibid., 377.

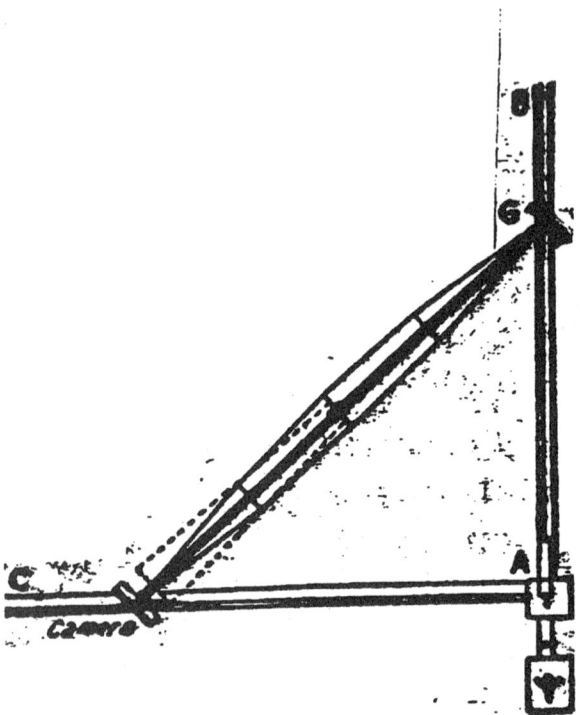

FIGURE 5. *Light entered this large spectroscope at A, struck the grating G, and was reflected to a plate holder serving as a camera to capture the resulting section of the spectrum.*

inches in vertical width. With a micrometer eyepiece in place of the camera box, the observer could view portions of the spectrum directly (thus changing the spectrograph into a spectroscope). Careful construction made the eyepiece micrometer "more like a dividing-engine than an ordinary micrometer." Here again, a precision screw made fine control possible; as in the ruling engine, the eyepiece screw was "to all purposes perfect."[110]

The most critical component in the whole spectrograph was the concave grating. With a ruled density of 10,000 to 20,000 lines per inch, and a typical ruled surface five and one-half inches long, the grating stood on two projections on a heavy platform of brass. The platform in turn stood on the carriage, which rested on the iron track, which rested on the wooden beam. While all other parts of the spectrograph were secured firmly, the grating was held, *"free from all constraint*, by a soft wax." The gentle grip of wax avoided any strain to the grating

---

[110]Ibid., 375, 377.

that might distort its shape. Here again, screws provided for adjustment, regulating the exact position of the frame supporting the grating.[111] The process of adjusting all the components of the Big Spectroscope could be lengthy. For example, Henry Crew once spent at least an entire day with Rowland on adjustments.[112]

The whole apparatus made possible a steady program of routine observation, in which Jewell played a significant part. Despite Rowland's hopes that the sale of gratings might pay for some of the work at Johns Hopkins on mapping the solar spectrum,[113] spectrographic work there in fact relied on the Rumford Fund of the American Academy of Arts and Sciences and the Bruce Fund of Harvard Astronomical Observatory. By 1891, Rowland reported, "The spectra of all known elements, with the exception of a few gaseous ones, or those too rare to be yet obtained, have been photographed in connection with the solar spectrum." Maps of some elemental spectra, as well as the solar spectrum, were being prepared. Research on the composition of the sun proceeded apace: "The greater part of the lines in the map of the solar spectrum have been identified and the substance producing them noted." But lurking among those Fraunhofer lines not yet identified might be some representing hitherto unknown elements.[114]

Rowland also further simplified the measuring process with his specially-built measuring engine, which enabled the measurer to read wavelengths directly from a scale. Thus the compilation of wavelengths became highly automated, and the possibility of human errors became greatly reduced. Yet Rowland's agency was still present, in the very machines he created, and the instruments gave him authority as well as renown. There were many competitors in the enterprise of establishing wavelengths, but Rowland believed his numbers "must still have a weight superior to most others published" because he used "such a powerful instrument" and because his laboratory took "such care in the determination of impurities."[115]

Rowland assumed that the comparability of solar and terrestrial spectra, as established by Kirchhoff and Bunsen, implied the comparability of the elemental compositions of the sun and Earth. The principal difference between matter in the two bodies, he believed, was temperature.[116] He often, although not exclusively, assigned the study of spectra from elements in the laboratory to his students, and largely kept for himself and Jewell the analysis of sunlight. Thus he held to

---

[111]Ibid., 375-377. Emphasis in original.
[112]Crew, 1886 Diary, 15 February, 18 February, HC.
[113]Rowland to Gilman, 25 September 1882, DCG.
[114]"Report of Progress in Spectrum Work," 521, 524.
[115]Ibid., 521, 523.
[116]Ibid., 523.

the celestial, and only relinquished some of the mundane. Overall, however, the work of the Hopkins school claimed both domains; in spectroscopy the two were rarely completely separate.

The Big Spectroscope remained in operation on the top floor of the physics department after Rowland. With it, the Rensselaer alumnus had solved many of the practical problems encountered in laboratory spectroscopy, and the survival of the big spectroscope meant the survival of those solutions, which remained available to numerous Hopkins investigators. For the master himself, knowledge of the construction and use of the overall apparatus remained a form of craft knowledge, which he de-emphasized in print in favor of the wavelengths he obtained. It remained for one of his students, Joseph Ames, to express the laboratory know-how explicitly, for later students and for investigators elsewhere.[117] Here again, it was thus the formation of a research school that made Rowland's experimental skill far more enduring than Fraunhofer's or Rutherfurd's.

---

[117] Ames, "The Concave Grating."

PLATE 1. Portrait of Professor Rowland by Thomas Eakins. Courtesy of the Addison Gallery of American Art, Phillips Academy, Andover Massachusetts

A School of Light

Veritas vos liberabit.[1]
- Motto of The Johns Hopkins University

High dispersion of the spectrum, which brought with it precise measurement, became a trademark and a special attainment of the Hopkins school. Joseph Ames not only made known the layout and workings of the Big Spectroscope; he also used it, and so held a privileged claim to authority, which he did not hesitate to use in print. After a Prague mathematician attempted a mathematical analysis of spectra in order to reason about the chemical elements, Ames criticized his data, and indeed much of the data obtained in spectroscopy thus far. He wrote, "The concave-grating gives the only accurate method of determining the ultra-violet wave-lengths of the elements; and, as a consequence of not using it, most of the tables of wave-lengths so far published are not of much value. So [Anton] Grünwald's error here may be great." The processes by which the chemical elements produced light remained unknown; formulas for the spectra of some elements meant only "that the molecules of those elements vibrate in general according to a similar law." In Ames's view, any attempt to theorize about the ultimate nature of matter could not succeed unless it encompassed the measurements that arose through the agency of the Hopkins school.[2]

---

[1]"The truth will make you free." John 8:32. Gilman probably chose this motto himself. French, *A History of the University*, 365-366.

[2]Ames, "Grünwald's Mathematical Spectrum Analysis," *Nature*, 40 (1889): 19.

In later studies, Ames stated the case for the new instrument even more strongly, and asserted that inferior instruments had vitiated many attempts to find numerical relationships between wavelengths of different spectra. One physicist, he wrote, had found mathematical terms relating some spectral lines, but the numbers were a product of chance, "for, using a prism spectroscope, his wave-lengths are arbitrary." Instrumental limitations also hobbled another physicist who attempted to assess the relations between elements and compounds: "The dispersion of his instrument was . . . so feeble that he could not properly decide upon similarities in different spectra." As for the hapless Grünwald, his physics may not have entailed any direct laboratory observation at all. Ames wrote, "It is evident from his writings that Prof. Grünwald has never had the benefit of using and working with a spectroscope."[3] For an American to reason without the personal benefit of empirical studies would have been unthinkable.

Ames wrote from a position of experimental strength. He had the benefit of a Rowland grating in the Big Spectroscope, of arc-spectra photographed by Rowland, and of wavelength values derived from Rowland's solar map—"as its superiority is now universally acknowledged."[4] He studied the spectra of zinc and cadmium, because they displayed a "most striking resemblance." Comparable series of lines, displaying perfect regularity, appeared in the spectra of both elements, and Ames wrote, "The more closely the two spectra are examined the more perfect does their agreement appear." In these two elements, lines in a series typically appeared as "triplets" (groups of three). The spectra may have held general significance, because the "series resemble greatly the series of hydrogen lines," and the spectra of many other elements included comparable series as well.[5]

Despite the detection of spectral patterns by Ames and many others, workable quantitative formulas were elusive: "For many years spectroscopists have sought to reduce the distribution of lines in any one spectrum to a mathematical formula, but without much success." Two approaches were possible. One entailed theoretical deduction and subsequent experimental test, while "the other is to give up all preconceived ideas and simply to 'guess,' as Kepler did." Because light entailed periodic motion, some investigators sought harmonic ratios between wavelengths to explain series relationships, but by 1890 such

---

[3]Ames, "On Relations between the Lines of various Spectra, with Special Reference to those of Cadmium and Zinc, and a Redetermination of their Wave-Lengths," *Philosophical Magazine* 30 (1890): 33, 36, 38; idem, "Some Notes on Spectra [sic] Analysis" (Ph.D. diss., Johns Hopkins, 1890); idem, "On some Gaseous Spectra:—Hydrogen, Nitrogen," *Philosophical Magazine*, 30 (1890): 48-58.

[4]Ames, "On Some Gaseous Spectra," 48-49; idem, "On Relations Between the Lines," 43.

[5]Ames, "On Relations Between the Lines," 45.

pursuits had largely ceased; the formulas that eventually prevailed were more complex. At the time of Ames's writing, a new formula by the Swedish physicist Johannes Rydberg held merit "in the division of the lines of the spectra into different series and in showing the relations between them." (The formula read:

$$n = n_0 - \frac{N_0}{(m + \mu)^2}$$

where n is wave number, n_0, N_0 and μ are constants, and m is a positive integer.) Although the formula did not express harmonic relationships, Ames still believed that spectra probably arose from "the fundamental or original vibrations of the 'molecule.'"[6]

Ames extended his own research to hydrogen, a spectrum "of unusual importance owing to its supposed presence in the light emitted by the white stars, and also because of Balmer's formula for the distribution of its lines which form a series." In the laboratory, Ames tried unsuccessfully to bring out only those hydrogen lines that occurred in a previously-known "stellar series"; in fact he captured them along with many other lines of hydrogen. Ames found a useful guide to the numerical results in the formula of J. J. Balmer, a Swiss teacher of mathematics:

$$\lambda = Cx\left(\frac{n^2}{n^2 - 2^2}\right)$$

where l is wavelength, C is a constant, and n is an integer. Balmer's formula was the first truly successful equation relating wavelengths in a spectral series, predating Rydberg's more complex formula. The earlier formula became an aid to discovery; Ames found that some lines of the Balmer series "were so faint as sometimes to escape notice until looked for." Thus he saw the lines only after the formula told him they might exist; simple mathematics led to the discovery of more facts. And Ames believed that the Balmer formula, with modifications, might apply to many or all spectra.[7]

For his measurements, he estimated his maximum error as 0.05 angstroms. After he separated out measurements of the stellar series, he belittled previous observations of that series by others: "It seems remarkable that the measurements of [physicist M.A.] Cornu and [astronomer William] Huggins should vary so much." Ames staked a claim to authority, based heavily on the apparatus at his command.[8]

Ames's work was exemplary, as was the course of his career. He became, successively, professor of physics (1891), director of the

---

[6] Ibid., 37, 40, 48.
[7] Ames, "On Some Gaseous Spectra," 51-52, 54-55.
[8] Ibid., 48-49, 54.

physical laboratory (1901), university provost (1926), and finally president of The Johns Hopkins University from 1929 to 1935. Indeed, he was the ultimate Hopkins man, because he retained affiliation with the university from the time he matriculated as an undergraduate in 1883 until his retirement as president. After receiving his bachelor's degree in 1886, he sailed for Europe with graduate student Henry Crew, who had been to Europe before. Ames enrolled at the University of Berlin for one year, but he returned afterward and became a laboratory assistant to Rowland, and received the Ph.D. in 1890.[9] After his appointment to the physics faculty in 1891, Ames "became the second man in the department and one upon whom Rowland relied and trusted absolutely, his *alter ego*, in fact."[10]

Furthermore, Ames's early work displayed attributes likely taken from Rowland: the deployment of practical solutions to the problem of measuring wavelengths; the reliance on comparisons of terrestrial and celestial spectra; the eagerness to conserve physics as he had learned it; and an international orientation. (Ames's publications by the early 1890s had addressed work conducted in Prague [Grünwald], Sweden [Rydberg], Switzerland [Balmer], France [Cornu], and England [Huggins].)

Yet there were also differences between Rowland, the first Hopkins physicist, and his "alter ego." One historian of the university observed that Ames "was not a brilliant investigator and facile inventor of new techniques and mechanisms, like Rowland; but was a better organizer and an exceptionally able teacher." He achieved this distinction despite a speech impediment, and valued "clear and exact speech" among students. On the grating, one chronicler wrote, "Ames was the most lucid expositor of the principles of its construction and use."[11]

The microculture of the school was not entirely technical. Ames did find ways to express material skills in words, making Rowland's tacit technical expertise explicit. And Rowland himself made also explicit statements of values that likely became tacit among his followers. Foremost among these was the laboratory ethos, which impelled the whole Johns Hopkins approach to experimental physics. Robert Kargon has described this ethic, which Rowland shared with Gilman. Both men "saw advanced education and research not merely as *training*, but as *mental and moral discipline*. It is impossible to overstress the *moral* component of their approach to science; it is but one of the

---

[9]John C. French, *A History of the University Founded by Johns Hopkins* (Baltimore: The Johns Hopkins University Press, 1946), 408.

[10]Crew, "Joseph S. Ames, 1864-1943," *BMNAS*, 23 (1945): 187. Emphasis in original.

[11]French, *A History of the University*, 408.

firmly-held viewpoints that separates their age from ours."[12] To Rowland, "the burden of imparting mental discipline to the student had shifted from the classics and moral philosophy to the natural sciences: the modern world demanded an educated leadership of 'doers.'"[13] This laboratory ethos, which Rowland expressed in formal speeches, rarely appeared later in the writings of Ames and other followers, who concentrated mainly on technical questions. But Rowland's normative statements continued to lurk somewhere in their past. At Hopkins, Kargon relates, "The rigor of precise measurement was . . . both a mental and a moral test."[14] The physicists became confident enough to express truth only when they became certain that their apparatus was working correctly.

Hesitancy was almost a virtue to Rowland. He stated, "For myself, I value in a scientific mind most of all that love of truth, that care in its pursuit and that humility of mind which makes the possibility of error always present more than any other quality." This valuation was almost paradoxical, for Rowland believed that the "scientific mind" would inevitably triumph; it was "destined to govern the world in the future and to solve problems pertaining to politics and humanity as well as to inanimate nature." Yet, despite this lofty destiny, the scientist had to be humble. The student should "be brought face to face with nature: let him exercise his reason with respect to the simplest physical phenomenon and then, in the laboratory, put his opinions to the test; the result is invariably humility." In nature, the scientist had to recognize truth.[15]

Existing historical accounts tend not to portray Rowland as humble, and there is ample evidence that he was not. When the Niagara Construction Company failed to pay his bill of $10,000 for consultation on a hydroelectric power project, he sued them, and testified in court. The lawyer representing Niagara Construction reportedly asked the physicist, "Who was the first physicist in the country?" To this question Rowland replied simply, "I am." The testimony clearly bolstered his bargaining position, and did not suggest humility. However, a colleague later remembered, "When we asked him afterwards how

---

[12]Kargon, "Henry Rowland and the Physics Discipline," 132. Emphasis in original. In nineteenth-century America, this ethic was not unique to Johns Hopkins. The Coast Survey, a federal agency which entered its heyday in the 1840s, conveyed the same ethic to scientific workers: "The practice of science on the Coast Survey was closely tied to the inculcation of moral character and discipline. Instrumentation encouraged disciplined habits of character necessary for the production of reliable knowledge." Hugh Richard Slotten, "The Dilemmas of Science in the United States: Alexander Dallas Bache and the U.S. Coast Survey," *Isis*, 84 (1993): 41.

[13]Kargon, *The Rise of Robert Millikan* (Ithaca: Cornell University Press, 1982), 32.

[14]Kargon, *The Rise of Robert Millikan*, 32.

[15]"The Physical Laboratory," 616, 618.

he could make such an answer, he replied that he was under oath and had to tell the truth."[16] (After a four-day trial, the jury awarded Rowland $9,000.)[17]

At the same time, humility was part of Rowland's culture. In 1874, when he was still a junior faculty member at Rensselaer, which celebrated its semi-centennial that year, the Rev. George N. Webber, doctor of divinity, delivered a sermon on "The Moral Influence of True Science" to a church packed with faculty, students, alumni, and friends of the Institute. Like Rowland's forebears in the clergy, Webber saw no inevitable conflict between religion and science; actually, the latter might deepen the former. The moral value of "true science" lay in its "influence on the character of the student." ("True science," in Webber's definition, would edify, while "false science" would merely give rise to vanity.)[18]

For Webber, four fruits of knowledge could be expected. Among them, *"The first fruit of true knowledge is humility."* Overweening pride might lead practitioners of false science "away from the modest and safe path of experiment, in which it has made such solid and brilliant achievements, into the fog-banks of *a priori* speculation." But students of true science should accept limitations.[19] Then came the second fruit of true science. Webber asserted, *"It encourages a religious spirit."* The student of nature would find ample reason for awe: "What can better inspire adoration than the study of nature? Not a fact discovered but opens a territory of limitless mystery to reverent faith. For every question which science answers, a hundred others start up which it cannot put to rest."[20] (Compare Rowland in 1883: "The field of research only opens wider and wider as we advance, and our minds are lost in wonder and astonishment at the grandeur and beauty unfolded before us.")[21] Thus science could deepen veneration: "As knowledge increases, nature becomes more and more an illuminated temple in which faith converses with and worships the Invisible."[22]

Thirdly, science should *"create a sense of the need of, and faith in, divine revelation."* Science did not necessarily undermine faith in the

---

[16]G. Stanley Hall, *Life and Confessions of a Psychologist* (New York: D. Appleton and Co., 1923), 237-238. A newspaper reported that the amount Rowland demanded was $30,000, or $150 per day. It said he was hired as one of the "highest scientific authorities." *Baltimore News*, 20 January 1894, in Manuscript Collection 1045, American Philosophical Society.

[17]Guralnick, "The American Scientist," 132.

[18]George N. Webber, "The Moral Influence of True Science," in *Proceedings of the Semi-Centennial Celebration of the Rensselaer Polytechnic Institute* (Troy, NY: Wm. H. Young, 1875), 9-11.

[19]Ibid., 12, 15-16. Emphasis in original.

[20]Ibid., 16-17. Emphasis in original.

[21]"A Plea for Pure Science," 613.

[22]Webber, "The Moral Influence," 20.

Bible, as long as the student regarded the revelations of the Bible as spiritual. Science still fostered a need for scriptural edification, because "the true scientific spirit . . . is sure to develop in the mind a sense of need of a supernatural enlightenment to answer those 'obstinate questionings' concerning the Unknowable which advancing knowledge of the cosmos perpetually conjures into consciousness, but can never answer."[23] Finally, science should foster charity: "Unless learning makes a man compassionate of the ignorance of others, and prompt to spread his knowledge . . . he surely knows nothing as he ought to know."[24]

After Rowland moved to Hopkins, his ministerial heritage did not make him unusual. Gilman remembered, "It was a curious fact, unobserved and perhaps unknown, that four of the first seven professors came from the families of gospel ministers."[25] With regard to religion, Rowland was "philosophic, not emotional,"[26] but Webber's address may have expressed an ethos of science proper to his era and culture. Along with humility, there was of course an admixture of pride. Instrument production was one source; idealism was another. Rowland believed scientists to be "a new variety of the human race." He wrote, "We form an aristocracy, not of wealth, not of pedigree, but of intellect and of ideals."[27] The cause of humility was not the scientist's fellow man, but nature. In teaching, Rowland truly had cause for humility; the crowning irony of the Hopkins school was Rowland's inadequacy in the classroom. He lectured four times a week, on subjects including electricity and magnetism, thermodynamics, and optics.[28] Yet one biographer remembered that he was "[u]nfit for the ordinary routine work of the class-room."[29] Remsen the chemist concurred to some extent: "His lectures were not eloquent. Words did not come freely to him. There was a lack of finish and elegance in his talks, but, on the other hand, he could say forcibly and clearly what he wanted to say."[30] Ames collaborated with Rowland on a physics textbook in the 1890s, and the experience prompted him to confide to Crew, "I don't think he knows anything about elementary teaching."[31] President Gilman went

---

[23]Ibid., 20-21. Emphasis in original.

[24]Ibid., 23.

[25]Gilman, *The Launching of a University*, 23.

[26]Remsen, "Henry Augustus Rowland," 425. Remsen also reported that the physicist "was as free from anything that could fairly be called sin as anyone I have ever known." Ibid., 426.

[27]"The Highest Aim of the Physicist" (1899), in *PPR*, 668.

[28]Kargon, "Henry Rowland and the Physics Discipline," 132.

[29]T.C. Mendenhall, "Henry Augustus Rowland," 133.

[30]Remsen, "Henry Augustus Rowland," 419.

[31]Ames to Crew, 18 June 1894, Henry Crew Papers - Microfilm (HCM), Niels Bohr Library, American Institute of Physics, College Park, MD. The textbook eventually appeared as Rowland and Ames, *Elements of Physics* (New York: 1900).

so far as to ask Rowland not to interfere with the undergraduate physics instruction given by a colleague, Charles Hastings.[32] John Miller has concluded, "Rowland spent most of his time during the seventies and eighties in the laboratory. His teaching received little attention and his lecturing was never regarded as particularly lucid."[33] However, these assessments are tempered by others. G. Stanley Hall later remembered of Rowland, "He was erratic and moody but supercharged with new ideas." The physicist may not always have remained close to his students, "but he always inspired them and they constituted a coterie by themselves. To hear him conduct an examination for the higher degree was an experience never to be forgotten."[34] Crew later remembered, "As a teacher he was inspiring to every serious minded student. Clear as a crystal. Confident of what he understood, but always so full of unsolved problems, that he was ever modest."[35]

How can Rowland's failings as a teacher be reconciled with the success of his research school? The answer is fairly simple: he taught "by example rather than by precept."[36] And he was an enabler; two generations of students profited by the Big Spectroscope. He taught far more effectively in the laboratory than the lecture hall. Crew wrote, "Of all the reforms in method the most revolutionary was the introduction of the student laboratory." This change spread rapidly: from about four in 1871, the number of "fairly equipped" teaching laboratories in the United States increased to about 400 by 1904.[37] Among the fourteen characteristics of a successful laboratory-based research school, as tentatively defined in the modern historiography of science, effective classroom teaching does not appear at all. Rather, the merits of a good research leader include: charisma, a strong research reputation, informality of approach, and institutional power.[38] These, and ten other characteristics, appeared to differing extents in the Hopkins School of Light, and an examination of these factors may help to explain Rowland's paradoxical success. The fourteen traits appear in Table 1.

Although the existence of factors one and five at the School of Light may be questioned or doubted, the remaining twelve almost

---

[32]Rowland to Gilman, 12 March 1884, DCG.

[33]Miller, dissertation, 289.

[34]Hall, *Life and Confessions*, 237. Hall's psychology laboratory lay within a physics building constructed in 1885.

[35]Crew to Robert S. Woodbury, 4 December 1935, HCM.

[36]T.C. Mendenhall, "Henry Augustus Rowland," 133.

[37]Crew, "Recent Advances in the Teaching of Physics," *Science*, 19 (1904): 485.

[38]The list of fourteen attributes appears in Geison, "Scientific Change," 24 table. It is echoed in Pamela M. Henson, "The Comstock Research School in Evolutionary Entomology," in *Osiris*, 8 (1993): 176 table. The traits are also applied in David Kushner, "Sir George Darwin and a British School of Geophysics," ibid., 220. Geison drew his list largely from J.B. Morrell, "The Chemist Breeders: The Research Schools of Liebig and Thomas Thomson," *Ambix*, 19 (1972): 1-46.

TABLE 1.
FACTORS AT WORK IN SUCCESSFUL RESEARCH SCHOOLS*
1. Charismatic leader
2. Leader with research reputation
3. Informal setting and leadership style
4. Leader with institutional power
5. Social cohesion, loyalty, esprit de corps, 'discipleship'
6. Focused research program
7. Simple and rapidly exploitable experimental techniques
8. Invasion of a new field of research
9. Pool of potential recruits (graduate students)
10. Access to or control of publication outlets
11. Students publish early under their own names
12. Produced and 'placed' significant number of students
13. Institutionalization in a university setting
14. Adequate financial support

*Drawn from Geison, "Scientific Change," 24

certainly obtained. The first four factors in Table 1 all directly concern leadership, reflecting what Geison continues to see as "the crucial role of individual human agency, personality and leadership in the fate of research schools."[39] Among these, "charisma" is the most difficult to define and assess, as it suggests personal qualities that may or may not be apparent in the historical record. J.B. Morrell, who introduced the notion of charisma to the historiography of research schools, described it as "extraordinarily effective, indeed messianic, leadership." He elaborated further:

> Such charisma, which was most effectively exerted in informal pre-bureaucratic contexts, helped to draw students in sufficient numbers to make the school viable. It enforced the standards and styles of work adopted by the school. It extracted from the students an unflagging almost fanatical devotion to research, particularly at times of intellectual failure and disappointment, and on occasion it also imposed fervent specialization.

> Further, the atmosphere at a research school "could be highly evangelical as the prophet broke through accepted conventions and led his devoted followers into unexplored and promising lands of enquiry."[40] (Consider again Rowland's almost-messianic tone in 1883: "Pure science is the pioneer who must not hover about cities and civilized countries, but must strike into unknown forests, and climb the hitherto inaccessible mountains which lead to and command a view of the promised land.")[41]

[39]Geison, "Research Schools and New Directions in the Historiography of Science," *Osiris,* 8 (1993): 235.

[40]Morrell, "The Chemist Breeders," 6-7.

[41]"A Plea for Pure Science," 608.

Many historical accounts until recently suggested that Rowland was devoid of anything that might be called "charisma." But historian Herbert Winnik has already taken issue with what he calls "the standard interpretation of Rowland as a one-sided personality with a cold remorseless devotion to science." Winnik differs with any suggestion that Rowland's "intellect was as poor in ideas, as his fancy in images and his soul in sensibilities." This and similar portrayals fit a stereotype too well. Rather, Winnik maintains, the physicist was "a complex man with intense beliefs." Nor did Rowland keep excessive distance from his students: "His method was to suggest a problem needing experimental investigation. He allowed the student to work out his own method of attack and the apparatus needed. He then criticized the results freely and frankly." This approach, Winnik suggests, could be inspiring.[42]

One historian of Johns Hopkins has alluded to a prevalent idea that Rowland neglected his students, suggesting that the picture may be oversimplified: "Persons who knew him well have said that this is not to be taken too literally; but the chief might well closet himself with his own research mechanisms while university fellows like Thomas Craig, [Arthur] Kimball, [Harry] Reid, and Ames carried on for him in the general laboratory."[43] Ames himself certainly had access to Rowland. He later remembered: "Those men who were his students would actually work in laboratory rooms with him or would be constantly consulting with him in regard to experimental difficulties and physical ideas."[44] The degree of access varied by individual.

Certainly Rowland did not exhibit charisma in the classroom. And he may have exercised little sway among the assembled faculty of Johns Hopkins, because he often missed faculty meetings.[45] But neither of these settings was the heart of his research school; rather, the laboratory was. Although the leader taught graduate students almost exclusively, one undergraduate encountered an activist in Rowland. "With his vigorous personality, keen pursuit of science, and incisive intelligence, Rowland provided an inspiring model for the budding young physicists."[46] The undergraduate, James Keeler, together with DeWitt Brace, recruited a new graduate student for the Hopkins school—Henry Crew, a recent graduate of Princeton. As Crew later recollected,

---

[42]Herbert C. Winnik, "A Reconsideration of Henry A. Rowland—the Man," *Annals of Science*, 29 (1972): 22, 23, 30, 31.

[43]French, *A History of the University*, 334-335.

[44]Ames, Speech, February 1935, JSA.

[45]Winnik, "A Reconsideration," 31.

[46]Osterbrock, *James E. Keeler*, 9.

In the autumn of 1883, I went to Berlin as a graduate student in Helmholtz's laboratory. There I fell in with James E. Keeler and D.B. Brace, each of whom had recently come from Rowland's laboratory. These two men told me so many stories about Rowland's fox hunting, boxing, sailing, working 36 hours at a stretch, making his own apparatus, etc. that my desire to know him was greatly stimulated. One day Brace remarked "Why, it is an inspiration simply to see that man walk along the corridor in his top-boots!" I could not resist the urge any longer. In a few weeks, I bade a reluctant goodbye to H. Kayser and to Helmholtz who gave me a nice letter to his old student, Rowland, and headed for Baltimore.[47]

There, although his laboratory work was very demanding, Crew was not disappointed by his first-hand experience with the leader. In August of 1884, he wrote of Rowland, "He makes things fly." Again, two days later, he wrote, "Rowland came out today + things moved fast." And some two weeks after that, he repeated the sentiment: "Prof. Rowland was out here + made things fly. I like him more, the more I see of him."[48]

Rowland's research reputation in the world physics community (the second factor at work in a successful research school) grew steadily over the course of his career. It began with the work on magnetism which the *American Journal of Science* initially rejected and which Maxwell, the Scottish physicist, praised. Rowland's reputation received the greatest single increase from his Berlin experiment on the magnetism of a rotating, electrified disk.

His reputation was also enhanced by his redetermination of the value of the ohm, which was important to international efforts to establish electrical standards. (The need for his intervention in international electromagnetic diplomacy was apparently quite strong. From Paris, he once wrote Gilman, "The Committee on electrical standards on which I am placed and which includes all the great physicists of Europe finds great difficulty in arriving at any conclusion.")[49] His redetermination of the mechanical equivalent of heat also augmented his reputation. In both the study of the ohm and the study of heat, precision was part of the claim to distinction. Then, in the 1880s and 1890s, Rowland led the way in high-precision spectroscopy, making himself and his measurements into examples not only for his own students, but for spectroscopists worldwide.

Ames attested to the importance of informal leadership in an informal setting (the third success factor). Even though he gradually rose

---

[47]Crew to Woodbury, 4 December 1935, HCM.
[48]Crew, 1884 Diary, 2 August, 4 August, 21 August, HC.
[49]Rowland to Gilman, 18 September 1881, DCG.

to the very top of the Johns Hopkins administration, he eventually looked back on the era of Rowland and Gilman as one that was happily free of rules and regulations. In his own day, Ames wrote, "The extent to which regulations, formal statements and credits prevail is alarming." Ames remembered a far different situation in the time of Gilman and Rowland, when there was "practically an entire absence of formal requirements." Uniformity and quantification of educational attainments did not yet prevail. Ames recalled that, to Gilman, "there was complete continuity between college and what we now call graduate work; he made no distinction between undergraduate and graduate. There were no formal requirements for admission to his new university." Study took priority over academic credits. Ames wrote, "The attainment of a degree was an incident, never an end to be aimed at." Reminiscing after the passage of forty years, Ames believed that the earlier situation was far more flexible: "There was no system of instruction, both professor and student had great opportunity to learn and to add to knowledge, the professors were the university."[50]

The leader's institutional power (the fourth factor) at Johns Hopkins was undoubtedly greater than it would have been elsewhere in the United States, because the new, well-endowed university endorsed the research interests of the faculty. The very newness of the situation brought academic freedom. Gilman recalled, "Every head of a department was allowed the utmost freedom in its development, subject only to such control as was necessary for harmonious co-operation. He could select his own assistants, choose his own books and apparatus, devise his own plans of study,—always provided that he worked in concord with his fellows." The university required only that each of the original six faculty discuss their plans with Gilman. Their freedom did not include financial management of the university, which was left in the hands of the trustees.[51]

In addition to the qualities of a research director, and the power given to him, the conditions for the success of a research school also included factors beyond the direct control of any one particular individual. A research director might interact well with each would-be physicist, but if that next generation of physicists did not interact among themselves, his success was limited. Thus social cohesion and a sense of shared interests (the fifth factor in Table 1) had to exist if the school was to hold a wider significance.

In the case of the Hopkins school of physics, we have the testimony of Hall that Rowland's students "constituted a coterie by

---

[50]Ames, "Recollections of a University Professor," *The Rice Institute Pamphlet,* 13 (1926): 211, 213, 215, 217-219.

[51]Gilman, *The Launching of a University,* 49-50.

themselves."[52] In a few cases, students began their research jointly. John Mohler and William Humphreys began research together in February 1895 on the effects of pressures greater than one atmosphere on the spectra of gases, eventually releasing measurements on twenty-three elements. Mohler then branched off alone to study the effects of very low pressures, and Humphreys studied alone the effects of high pressure on additional elements.[53]

Another two students both entered the school in 1895, and jointly finished a thesis on "The Radiation of a Black Body," a subject of great interest to theoretical physicists in Europe at that time. Charles E. Mendenhall began the work with the help of Harry Reid, who had received his Hopkins Ph.D. in 1885 and stayed to teach. Mendenhall observed thermal spectra from a black body at temperatures above 500° C. The second student, Frederick A. Saunders, then conducted studies below that temperature. Their apparatus resembled that which Langley had used to measure energy across the solar spectrum.[54]

Apart from the Humphreys-Mohler and Mendenhall-Saunders collaborations, however, doctoral research was apparently very much each student's own. Finished reports typically expressed a debt only to Rowland and/or Ames, and sometimes to Lewis Jewell. James Barnes was thus fairly unusual in expressing thanks to his "fellow students, whose kind assistance in word and deed has greatly facilitated these experiments."[55]

Outside the laboratory, students were involved in journal meetings, to discuss current developments in physics.[56] They could also have benefitted from a larger social context in which Rowland's laboratory "had become one of the crossroads of the scientific world, through which scientific men from all over the world passed, and where they were prone to linger." The students did not necessarily neglect each other for the admiration of great men, however, for among themselves "there were younger men who in the next few years were to remake the place of physics in American institutions. Among these men friendships of lifelong value were to be made."[57] In addition, a

[52]Hall, *Life and Confessions*, 237.

[53]John Fred Mohler, "Studies in Spectrum Analysis" (Ph.D. diss., Johns Hopkins, 1895), 34. W.J. Humphreys, "Changes in the Wave-Frequencies of the Lines of Emission Spectra of Elements, their Dependence upon the Elements themselves and upon the Physical Conditions under which they are Produced" (Ph.D. diss., Johns Hopkins, 1897), 230.

[54]Mendenhall and Saunders, "The Radiation of a Black Body," 32. Reprinted from *ApJ* 13 (1901): 25-47.

[55]James Barnes, "On the Analysis of Bright Spectrum Lines" (Ph.D. diss., Johns Hopkins, 1904), 211. Reprinted from *ApJ*, 19 (1904): 190-211.

[56]On the journal meetings, see Crew, 1884 Diary, 29 October; 1886 Diary, 27 January, HC.

[57]A.A. Knowlton, "Henry Crew (1859-1953)," *Isis*, 45 (1954): 171-172.

Science Association at the university, involving both faculty and students, allowed for further discussion, outside the laboratory, of major developments in science.[58]

The program of research (factor six) shared by Hopkins spectroscopists was quite literally focused, by the curved surface of the concave grating. More conceptually, it remained focused on the comparison of solar and terrestrial spectra. It was conducted by means of the "simple and rapidly exploitable experimental techniques" (factor seven) associated with the utilization of a Rowland grating, often in the Big Spectroscope. Students in the School of Light typically used Rowland instruments to study spectra of the elements, especially under varying conditions of temperature or pressure, with reference to lines in the solar spectrum. Crew was among the first students to use the spectroscopic apparatus, when he investigated the rotation of the sun, based on the apparent Doppler shifting of spectral lines from either side of the solar disc.[59] However, his thesis was not typical, and in fact his observations were eventually disregarded.[60]

A more representative early student was Ames, who studied the laboratory spectra of cadmium, zinc, hydrogen, and nitrogen. He studied hydrogen and nitrogen at the suggestion of Rowland, and used the instrument "mounted in the well-known Rowland manner." (Indeed, it was Ames himself who had made the manner well-known.) Ames's experiments were also "conducted according to the method designed by Prof. Rowland," and owed various and sundry other debts to the leader, as well. Using an induction coil connected to aluminum electrodes in a tube containing the gases, Ames obtained a brilliant discharge. Photographing the gaseous spectra required exposures of as much as one hour. In his published report, Ames gave his wavelength measurements, compared them with measurements by others elsewhere, and explored their possible significance.[61]

John Mohler, who had taught mathematics and science in public schools prior to his arrival at Hopkins, initiated work together with William Humphreys, as mentioned above. Humphreys was likewise an erstwhile teacher of mathematics and science. Starting from their observations of gases at various pressures in the laboratory, they worked together, and in consultation with Lewis Jewell, in evaluating the pressure in the solar "reversing layer"—where the absorption of

---

[58]Crew, 1884 Diary, 5 November; 1886 Diary, 3 February, HC.

[59]Crew, "On the Period of Rotation of the Sun as Determined by the Spectroscope" (Johns Hopkins, Ph.D. diss., 1887). In *American Journal of Science*, 135 (1888): 151-159. Crew, 1886 Diary, 24 November; 1887 Diary, 7 January, 17 February, 8 March, 30 March, 6 April, 9 April, HC.

[60]"Crew later learned that his measurements were without value." Weart, "Crew, Henry," *DSB*, Vol. 17 (1990), ed. Frederic L. Holmes, 189.

[61]Ames, "On Some Gaseous Spectra," 48-49, 51, passim.

sunlight from lower layers caused the dark Fraunhofer lines in the spectrum. Mohler concluded that pressures in the reversing layer varied by element, ranging from two atmospheres for aluminum, to seven atmospheres for iron.[62]

Mohler's principal accomplishment under his name alone concerned gases at low pressures (less than one atmosphere). Rowland and Ames directed his work, and supplied tools: "The pressure apparatus was made by Professor Rowland several years ago for this purpose but laid aside for more important work." A.A. Michelson sent Mohler a tube of cadmium similar to the tube that the former had used to determine the length of the standard meter bar in Paris, and another physicist loaned a tube of helium that was first filled by the physicist who discovered the element, William Ramsay. The Rowland grating had a focus of 21 _ feet, and the light source, an electric arc, drew five to fifty amps at 110 volts direct current, producing a two-inch spark. Precision required quietude, and thus most of the work "was done at night when the Physical Laboratory is comparatively free from vibration and when consequently one can obtain better definition than at other times."[63] During the day, traffic outside could cause minute tremors that would reduce the precision of observations.

The solar-terrestrial comparison was implicit in the observations, because Mohler recorded both arc lines and solar lines on the same photographic plates. He hoped to account for discrepancies between the wavelengths of solar lines and the wavelengths of corresponding laboratory arc lines, both as determined by Jewell and Rowland. The two had expected exact coincidence, but had not always found it. Mohler's estimated accuracy for his own measurements was about three thousandths of an angstrom, and he exposed more than 140 negatives altogether. He concluded that spectral lines shift toward the red with increasing gas pressure, and toward the violet with decreasing pressure. The effect was approximately proportional to wavelength.[64] The conservative cast of Hopkins research was evident in Mohler's efforts to reconcile expectations with observations. Rather than give up the comparability of solar and terrestrial light, he hoped to keep the two types of spectra interpretable in terms of each other, by evaluating pressure as a complicating factor.

---

[62]Mohler, "Studies in Spectrum Analysis," 34, 35, 29. Mohler published separately some of his joint studies at about the same time. Humphreys and Mohler, "Effect of Pressure on the Wave-Lengths of Lines in the Arc-Spectra of Certain Elements," *ApJ* 3, (1896): 114-137. L.E. Jewell, Mohler, and Humphreys, "Note on the Pressure of the 'Reversing Layer' of the Solar Atmosphere," ibid., 138-140. Mohler and Jewell, "On the Wave-Length of Some of the Helium Lines in the Vacuum Tube and of D_3 in the Sun," ibid., 351-355.

[63]Mohler, "Studies in Spectrum Analysis," 2-3, 30, 34-35.

[64]Mohler, "Studies in Spectrum Analysis," 1, 5, 6-8.

Almost all student research on spectra used techniques that relied upon existing apparatus, and very often the solar tables were a guide. Other investigations that utilized a Rowland grating, along with various other apparatus constructed by the leader, concerned: the spectrum of mercury, which was often used in laboratory pumps, and so might contaminate samples;[65] the effect of a magnetic field on the spectrum of iron;[66] the accurate measurement and identification of spectral lines in the little-known infrared regions of the spectrum;[67] a classification, and attempted explanation, of the appearance of lines in the spark spectrum of cadmium;[68] and the tabulation of lines from certain relatively rare elements.[69] The two authors of the rare-element study indicated that it "may be regarded as a continuation of the work upon the solar spectrum commenced several years ago by one of us, inasmuch as the ultimate object in view is the identification of some of the many lines in the spectrum of the sun, whose origin still remains unknown."[70] The school kept producing Rowland-style spectrum studies into the 1920s.

The field of research that the School of Light invaded (factor eight) was quite new. Although the first glimmerings of modern spectroscopy arose from Fraunhofer's work, the guiding principles of the discipline had only become fully clear in 1859 and after. Accurate quantitative analysis came still later; according to one authority on spectroscopy, Ångström's engraved map of the solar spectrum of 1868 "marked its birth as an exact physical science." Thus when Rowland made his invention in 1882, spectroscopy had only existed for some fourteen years as an exact science. "More and more experimenters entered this field of research, and advances were made in every direction."[71]

Ångström's map did not prove to be a final statement; Rowland's map showed the actual wavelengths of most lines to be greater than those stated by Ångström, by an amount varying between 0.5 and 1.8

---

[65]William B. Huff, "The Spectra of Mercury" (Ph.D. diss., Johns Hopkins, 1900). Reprinted from *ApJ*, 12 (1900): 103-119.

[66]N.A. Kent, "Notes on the Zeeman Effect" (Ph.D. diss., Johns Hopkins, 1901). Reprinted from *ApJ*, 13 (1901): 289-319.

[67]Exum Percival Lewis, "The Measurement of Some Standard Wave-Lengths in the Infra-Red Spectra of the Elements" (Ph.D. diss., Johns Hopkins, 1895). Reprinted from *ApJ*, 2 (1895): 1-25.

[68]Charles Carroll Schenck, "Some Properties of the Electric Spark and Its Spectrum" (Ph.D. diss., Johns Hopkins, 1901). Reprinted from *ApJ*, 14 (1901): 5-24.

[69]Robert R. Tatnall, "The Arc-Spectra of the Elements: Boron, Beryllium, Germanium, Platinum, Osmium, Rhodium, Ruthenium and Palladium" (Ph.D. diss., Johns Hopkins, 1895). This research also appeared in installments in print, under joint authorship: Tatnall and Rowland, "The Arc-Spectra of the Elements. I. Boron and Beryllium," *ApJ* 1 (1895): 14-17; "II. Germanium," ibid., 149-153; "III. Platinum and Osmium," ibid., 2 (1895): 184-187; "IV. Rhodium, Ruthenium and Palladium," ibid., 3 (1896): 286-291.

[70]Tatnall and Rowland, "The Arc-Spectra of the Elements. I. Boron and Beryllium," 14.

[71]E.C.C. Baly, *Spectroscopy* (London: Longmans, Green, and Co., 1912), 33.

angstroms. "By common consent, in view of the accuracy of Row-land's method and work, his scale was universally adopted as the standard of reference."[72] Thus the Hopkins school seized ground that was still highly contested, where it could still make important claims to authority. Definitive spectra of many elements, containing spectral lines that might match lines in the solar spectrum, did not exist, and promised a fertile field for the school.

It was important to the Hopkins school that adequate numbers of future spectroscopists be recruited (factor nine). As Morrell observed, "Quite simply there had to be a regular supply of motivated students who were keen to apprentice themselves to a recognized or emerging master of his subject."[73] This supply emerged; as noted in Chapter 1, there was an abundance of students in Rowland's laboratory, even though many did not take degrees. For many, education was their principal profession, both before and after graduate study. Among those who did complete degrees in the School of Light through 1909, many had prior experience teaching science or mathematics in sec-ondary schools.[74] Some came directly from college,[75] and some had college teaching experience.[76]

One of the most notable recruits to the School of Light, Robert Wood, came the short distance from the chemistry department at Johns Hopkins where, Dieke notes, "He found that he was more in-terested in what went on in Rowland's laboratory in the Department of Physics than what he was supposed to be doing in the Chemistry Department."[77] It was already clear that spectroscopy provided a very sensitive means of performing chemical analysis, if the spectra of the elements being analyzed were known. Wood's best-known accom-plishment as a student was, though anecdotal, also highly illustrative of the power of spectroscopy for chemical analysis. At the boarding house where Wood lodged, some of the tenants suspected that the landlady used scraps from their dinner plates as ingredients for the breakfast hash of the next morning; after a supper of steak, hash al-most invariably appeared at breakfast. One evening, Wood left some choice scraps of meat on his plate, and sprinkled them with lithium chloride. The following morning, he took a quantity of the breakfast

---

[72]Ibid., 34.

[73]Morrell, "The Chemist Breeders," 4.

[74]They included Frank Cooper, T. Sidney Elston, William B. Huff, Humphreys, Clinton Kilby, and Mohler.

[75]Including James Barnes, David Guthrie, and August Pfund.

[76]Including Humphreys, Charles E. Mendenhall, James Porter, Harvey Rentschler, and Harry Springsteen.

[77]G.H. Dieke, "Robert Williams Wood," *Biographical Memoirs of Fellows of the Royal Society* 2 (1956): 328.

hash to Rowland's laboratory, burned it in a flame, and captured the light with a spectroscope. Unmistakably, a crimson spectral line distinct to lithium appeared, showing that supper scraps had indeed been converted to breakfast hash.[78] Wood's longer career was rife with anecdotal encounters such as this.

After Rowland's death in 1901, Wood became his successor as professor of experimental physics at Johns Hopkins. He did not eclipse the field as Rowland had, but he was nevertheless a memorable figure. A biographer noted, "Wood never formed a school; he was too individualistic for that. He had, however, many students from all over the world and even now whenever two of them come together they can spend hours telling anecdotes about their former teacher."[79] He also wrote an important textbook on optics.[80]

The true success of a research school required not only that students arrive in number, but that many subsequently achieve fame. Morrell has recognized, in his study of the discipline of chemistry, that "relatively easy access to publication opportunities, or best of all control of them, enabled a school to convert private work into public knowledge and fame."[81] (Access to publication appears in Table 1 as factor ten.) In the case of the School of Light, the most important avenue of publication was the *Astrophysical Journal*, founded in 1895. A central tenet of the Hopkins school, the comparability of celestial light with laboratory light, also guided the newly emerging discipline of astrophysics, and the editorial policies of the *Astrophysical Journal*. Thus, although Rowland's school has never been described as a breeding ground of astrophysics, the new discipline was in fact indebted to the many physics studies from Hopkins that appeared in the Chicago-based journal. (In full title, the periodical was called *The Astrophysical Journal: An International Review of Spectroscopy and Astronomical Physics*.)[82]

The principal founder of the journal was George Ellery Hale. He had been launched into a career in solar spectroscopy in part by the acquisition of a Rowland grating at the age of seventeen. While a junior at the Massachusetts Institute of Technology, Hale had talked at length with Rowland.[83] After earning the B.S. in physics at MIT in

---

[78]William Seabrook, *Doctor Wood: Modern Wizard of the Laboratory* (New York: Harcourt, Brace and Co., 1941), 39-40.

[79]Dieke, "Robert Williams Wood," 328, 333.

[80]Wood, *Physical Optics* (New York: Macmillan Co., 1905). A second edition appeared in 1911, and it was reprinted in 1924.

[81]Morrell, "The Chemist Breeders," 5.

[82]*ApJ*, 1 (1895): title page.

[83]Hale to Harry Manley Goodwin, 6 June 1889, George Ellery Hale Collection (GEHC), Huntington Library, San Marino, CA.

1890, Hale considered graduate study under Rowland at Hopkins,[84] before deciding against graduate study altogether, in favor of private research at his home near Chicago. (In 1893, he did study in Berlin under Helmholtz and Max Planck.)[85] Hale's co-editor was James Keeler, who completed his undergraduate studies at Hopkins only five years after it opened.[86] The five assistant editors included Joseph Ames and Henry Crew, and the ten associate editors included Rowland himself. Thus the Hopkins school not only had access to the journal, but had significant control, as well.

Of sixteen dissertations on spectroscopy and light completed under Rowland, eleven were published in some form in the *Astrophysical Journal*.[87] No other channel of publication even approached this figure. Two studies appeared in the *American Journal of Science*,[88] and one, Ames's thesis, appeared in two parts in the British *Philosophical Magazine*. In the quarter century after Rowland, when the Hopkins school of physics produced forty-four dissertations on spectra, George Ellery Hale's journal maintained its position as the leading vehicle for the school. Twenty-six of the doctoral studies appeared in the *Astrophysical Journal*. The only other journal carrying a significant number was *The Physical Review*, with nine.[89] Thus, in its earliest origins, the American physics discipline held a strong affiliation with astrophysics, specifically in the tenet, common to both, of the comparability of celestial and terrestrial light.

Although some articles appeared under the joint authorship of Rowland and a student, or later of Wood and a student, for the most part students in the Hopkins School of Light published under their own names. In some cases their research appeared in print even before it appeared as a formal dissertation. Thus success factor eleven, students publishing early under their own names, was indeed evident for the School of Light.

As the completion of a thesis constituted the production of a Ph.D., the school also produced a significant number of spectroscopists: sixty Ph.D.s over the first fifty years of the department. (Their names are listed in Table 2.) However, true success, and a selfperpetuating American discipline of spectroscopy, required not only the conferral of degrees, but also the successful placement of spectroscopists in

---

[84]George Ellery Hale to Rowland, 3 July 1890, HAR.

[85]Wright, "Hale, George Ellery," *DSB*, Vol. 6 (1972), 28.

[86]Osterbrock, *James E. Keeler*, 14.

[87]The joint dissertation of C.E. Mendenhall and Frederick Saunders is counted twice in these totals.

[88]Crew's thesis, and Louis Bell, "The Absolute Wave Length of Light," *American Journal of Science*, 135 (1888): 265-282.

[89]"Doctors' Dissertations," 41-47.

TABLE 2.		
PH.D.s IN SPECTROSCOPY AND LIGHT		
1876–1901		
Joseph Ames	William Jacques	Harry Reid
Louis Bell	Norton Kent	Frederick Saunders
Henry Crew	Exum Lewis	Charles Schenck
Caleb Harrison	Charles Mendenhall	Robert Tatnall
William Huff	John Mohler	
William Humphreys	Herbert Reese	
1901–1926		
John Anderson	Robert Dickey	Felix Hackett
James Barnes	Charles K. Edmunds	Janet Tucker Howell
Frederick Brackett	Alexander Ellett	Edward Hulburt
Taylor Carter	Thomas Elston	Herbert Ives
Robert Castleman	Mabel Frehafer	Enoch Karrer
George Clinkscales	Rogers Galt	Clinton Kilby
Frank Cooper	Philip Gottling	
Charles Deppermann	David Guthrie	

professional positions (factor twelve). Students did not truly enter the profession until they became accepted as working physicists. Although the School of Light produced many more doctors of physics in the second quarter-century than in the first, the placement rate was much higher during the earlier period. Of the sixteen students who completed their studies during Rowland's era, fifteen were still listed as active physicists in a 1906 directory. The great majority taught physics at colleges or universities. One listed his position as professor of physics at the U.S. Weather Bureau, and another was a consulting engineer who lectured part-time on electrical power.[90]

Contrasted with the ninety-four percent placement rate achieved by 1906, the placement rate at the close of the second quarter-century was much lower, only sixty-four percent. Of the forty-four spectroscopists trained between 1902 and 1926, only twenty-eight were listed as active physicists in a 1927 directory. For those who remained in the physics profession, the pattern of employment differed from the earlier period, in that research positions were much more common by the 1920s, accounting for thirty-six percent of the physicists who had emerged from the Hopkins laboratories after 1901, with the remainder still teaching college or university physics. One federal institution, the National Bureau of Standards, employed three alumni of the School of Light as of 1927. By then, the Bureau had employed eight of the second

---

[90]Information from *American Men of Science: A Biographical Directory*, ed. J. McKeen Cattell (New York: The Science Press, 1906; first edition).

generation of Ph.D.s at one time or another.[91] Although the attrition rate at the Hopkins school by the end of the second quarter-century, thirty-six percent, was much greater than the attrition rate by the end of the first quarter-century, six percent, the School of Light still produced a total of forty-three working physicists over the two periods.

Universities emerged apace in the U.S. in the late nineteenth and twentieth centuries, but in 1876 the university setting (factor thirteen) of the School of Light was something new in the United States. The practices and philosophy of The Johns Hopkins University did much to establish the subsequent meaning of the word "university" in American life. Although Baltimore merchant Johns Hopkins provided in his will for the endowment of the new university, it was Gilman, the first president, who suggested to the trustees "that they should give emphasis to the word 'university' and should endeavour to build up an institution quite different from a 'college,' thus making an addition to American education, not introducing a rival."[92]

Another innovation of the new university was its secular character. Gilman wrote, "The public had been so wonted to regard colleges as religious foundations, and so used to their control by ministers, that it was not easy to accept at once the idea of an undenominational foundation controlled by laymen." This system of governance did not make the university an irreligious place; Gilman, the initial moving force at Johns Hopkins, believed that "science is the discoverer and interpreter of . . . divine order."[93]

Johns Hopkins became an ideal setting for the cultivation of research schools. Ira Remsen, the head of the chemistry department, contributed even more Ph.D.s to American chemistry than Rowland contributed to American physics. His view of the moral function of laboratory instruction was much like Rowland's and Gilman's. As Owen Hannaway has observed, "Remsen's laboratory was as much a place for instilling the virtues of work and discipline as it was a nursery of new knowledge." But unlike Rowland's students, Remsen's students "remembered him best as a skilled lecturer and influential teacher, not as a dynamic director of research."[94] Yet Remsen and Rowland both placed the ethical status of science on a footing almost equal to that of religion. Remsen wrote in 1903, "A life spent in accordance

---

[91]Information on 1902-1926 Ph.D.s drawn from *American Men of Science: A Biographical Directory*, eds. J. McKeen Cattell and Jaques Cattell (New York: The Science Press, 1927; fourth edition). As regards their selection criteria for the listings, the editors stated, "Efforts have been made to include all living Americans who have contributed to the advancement of science." The requirements were about the same as those for membership in a scientific society. viii.

[92]Gilman, *The Launching of a University*, 7-8.

[93]Ibid., 23, 251.

[94]Hannaway, "The German Model of Chemical Education," 145, 153.

with scientific teachings would be of a high order. It would practically conform to the teachings of the highest type of religions."[95]

In biology, Irish-born Henry Newell Martin came from Cambridge University, acquainted with "all the newer methods of physiological research." Gilman hired him for the laboratory teaching methods that Martin and his mentors had employed in England. Gilman later recalled that, because plans called for the eventual creation of a medical school, "it was clear that a preparatory study of the biological sciences should be encouraged by methods superior to any which were then employed in this country, and of far greater comprehensiveness." In England, Martin was not only an associate in the research school of Cambridge physiologist Michael Foster, but also an assistant of T.H. Huxley. Some Americans, however, feared the importation of biology to the U.S. Gilman said, "The science was dreaded as if it were to overthrow, or at least to undermine, religious belief."[96]

With human health at stake in the conception of a medical school, however, biology, or at least the subspecialty of physiology, was required. Gilman said, "It was soon determined that no one should be encouraged to enter upon the study of medicine without a careful previous training in a physiological laboratory. The improvements now common in medical schools are largely based upon the recognition of the principle that living creatures, in their normal and healthy aspects, should be studied before the phenomena and treatment of disease."[97] Like Remsen and Rowland, Martin experienced success. Laboratory instruction under Martin made Johns Hopkins "the leading training ground for the next generation of American biologists and physiologists."[98]

Laboratory instruction, of a sort, even arose in mathematics. Like Martin, James Joseph Sylvester was called from England to form a department. Sylvester hoped to establish a self-perpetuating school of mathematical investigators, and achieved some limited success through establishment of a "mathematical seminarium." Karen Hunger Parshall writes:

> A kind of laboratory along the lines of a physical sciences laboratory, the mathematical seminarium consisted of Sylvester, the director, and his auditors, the laboratory assistants. Under the director's guidance, the laboratory staff as a whole studied the various aspects of a particular problem at hand.

---

[95]Ira Remsen, "Scientific Investigation and Progress," *Proceedings of the American Association for the Advancement of Science*, 53 (1904): 14.

[96]Gilman, *The Launching of a University*, 51-52.

[97]Ibid., 52.

[98]Geison, *Michael Foster and the Cambridge School of Physiology*, 142.

But Sylvester left Johns Hopkins for Oxford in 1883, and his school "failed to go out from Baltimore and create enclaves of pure mathematical research around the country."[99] Thus the success of the research school in mathematics was short-lived, unlike the schools in physics, chemistry, and biology.

While the human agency of a director was crucial to each research school, the simultaneous emergence of four research schools at Johns Hopkins demonstrated the possibilities that could arise within a favorable institutional setting. The four research schools required not only exceptional individual leaders, but exceptional conditions, as well. Like the combination of diffraction grating and research school in physics, the combination of activist research directors and supportive university administration brought about a synergetic effect that neither factor alone could have produced.

Of course, a university could not exist without some form of financial support (factor fourteen), and the endowment bequeathed in 1873 by Johns Hopkins, a Quaker, founded the new institution. The bequest of three and a half million dollars was the largest single gift ever made, until then, to an American institution of higher education.[100]

The willingness of the university to spend more than $6,000 for physical apparatus, at Rowland's request, followed from the size of its endowment. For the study of light, by 1879 the new physics laboratory possessed two German-made spectrometers, one German-made prism spectroscope, a heliostat (for directing sunlight to a fixed observing position), more than a dozen prisms, and a camera and darkroom for conducting photography. It also had three small, American-made diffraction gratings ruled on Lewis Rutherfurd's engine, and the respective solar maps of Gustav Kirchhoff, Anders Ångström, and Rutherfurd.[101]

The university did not skimp on Rowland's salary, either. In December 1875, before the institution had even opened, it raised his salary from $1,600 to $2,000, and when Rowland became full professor in April 1876, his salary increased again, to $3,000.[102] Despite this largesse, the initial conditions at the physics department were austere. Crew recalled, "When I went to Baltimore he was lecturing in an

---

[99]Parshall, "America's First School of Mathematical Research," 166-167, 168, 171, 191. After Sylvester, Simon Newcomb directed the mathematics department, from 1883 to 1893, serving as professor of both mathematics and astronomy. Newcomb also worked for the United States Navy, and kept this affiliation during his term at Hopkins. French, *A History of the University Founded by Johns Hopkins*, 140.

[100]Hugh Hawkins, *Pioneer: A History of the Johns Hopkins University, 1874-1889* (Ithaca: Cornell University Press, 1960), 3-4.

[101]"List of Apparatus," *Harvard University* [Library] *Bulletin*, no. 12 (1879): 351-352.

[102]Miller, dissertation, 184, 190.

upstairs bed-room of what was formerly a private house. His office was in a rear part of the house, in a sleeping room over the kitchen."[103] To ameliorate this situation, in 1885, the university constructed a new building expressly designed by Rowland himself to serve as the new physics laboratory. The four-story building was the largest constructed thus far by Johns Hopkins, covering about seventy feet by 115 feet and costing $175,000. The new building also included an astronomical observatory, lecture rooms for mathematics, and a laboratory for psychologist G. Stanley Hall as well as a machine shop and a dynamo room.[104] The new structure was the most substantial and the most concrete (actually red brick and sandstone)[105] expression of the financial support given to Rowland.

Of course, Rowland did not have *carte blanche* throughout his entire quarter-century at the university. Both Rowland and Ames felt by 1897 that their department had fallen behind in the creation and procurement of necessary equipment. Ames argued to Gilman, "New apparatus is constantly needed, which we either buy or do without. A great deal of this could be done by a trained machinist." Even the upkeep of existing apparatus required more hands than Schneider's: "Repairs are needed every day of the year, and needed instantly."[106] Where once the Hopkins physics department was among the best-equipped in the United States, Rowland in 1897 felt that it no longer was, and that the university should hire another machinist. He wrote, "In order for our laboratory to be on an equal footing with those of Chicago, Cornell, Columbia, Amherst and others such a step is absolutely necessary. All the institutions named employ from two to five mechanics making and repairing apparatus."[107] However, in the crucial first decade or so of existence, from 1876, the department had adequate shop facilities, and indeed led the way in the discipline.

Of all the fourteen success factors examined in the case of the School of Light, only two are in any doubt: Rowland's charisma (one) and the social cohesion of the students (five). These are qualities which are difficult to assess historically, without opportunity for direct observation. Certainly the historical evidence does not rule out the presence of these two factors, and it suggests that they may well have existed. In setting forth the categories in Table 1, Geison acknowledged that they may seem vague and arbitrary, but also suggested that they could

---

[103]Crew to Woodbury, 4 December 1935, HCM.

[104]Kargon, "Henry Rowland and the Physics Discipline," 133. French, *A History of the University,* 61 and facing page.

[105]French, *A History of the University,* 62.

[106]Ames to Gilman, 5 February 1897, Rowland-Ames correspondence, HAR.

[107]Rowland to Gilman, 13 March 1897, Rowland-Ames correspondence, HAR.

serve as "a rough guide to the sorts of factors to be considered by students of research schools."[108] My conception of the Hopkins School of Light, and the fourteen-factor historiographical model, tend to validate each other. The analytical framework of the model offers a means of understanding how science as a profession is propagated, and in this instance how an American profession of physics came into being.

However, most of these factors were internal to the university, or at least to the profession. They would not necessarily have been immediately obvious to contemporaries of Rowland or Ames; here, they have been reconstructed from historical sources. The principal means by which the work of the School of Light became known was the *Astrophysical Journal*, chiefly founded by a solar spectroscopist who was, as it were, a satellite of the school, having utilized Hopkins instruments, conferred deeply with the Hopkins leader, and pursued the Hopkins style of research in his own distant observatory. The internal workings of the school were less manifest. Although George Ellery Hale, principal founder of the *Astrophysical Journal*, decided against graduate study in the Hopkins School of Light, he became far more valuable to the school and to its research program in Chicago than he ever could have been in Baltimore. Thus a consideration of his professional development is relevant.

---

[108]Geison, "Scientific Change," 26-27.

CHAPTER

The Physics of Heaven and Earth

Rowland, if report be true, declared himself under oath to
be a physicist, but his contributions to astronomical
spectroscopy were very great.
- Henry Norris Russell, 1947[1]

Even where the school did not reach, Rowland's instrument alone promoted high-dispersion, high-precision spectroscopy. Physicists at many locations benefitted from the material embodiment of Rowland's expertise, and thus pursued research of the kind and caliber carried out in Baltimore. Whether in Germany, France, England, or other parts of the United States, physical laboratories adopted Hopkins-style spectroscopy when they made use of the concave diffraction grating. But while the recipients of Rowland instruments numbered in the hundreds or more, the instrument by itself had the greatest effect on George Ellery Hale.

Hale first received a Hopkins-made diffraction grating when he was a boy of seventeen in suburban Chicago; he purchased a flat grating, one inch across, through Brashear in Pittsburgh.[2] In Helen Wright's words, "He saw that, like a vast suspension bridge, the

---

[1]Russell, "America's Role in the Development of Astronomy," *Proceedings of the American Philosophical Society* 1947 (91): 14.

[2]Helen Wright, *Explorer of the Universe* (New York: American Institute of Physics, 1994; first published 1966), 43. Karl Hufbauer, *Exploring the Sun: Solar Science since Galileo* (Baltimore: The Johns Hopkins University Press, 1991), 71.

spectroscope could help him travel the millions of miles from earth to sun, and so out to distant stars and nebulae. With it he could bring them into the laboratory to be analyzed as surely as any earthly substance." At the higher dispersion made possible by the Rowland plane grating, Hale began measuring wavelengths of solar absorption lines in 1886.[3]

Like the elder Rowlands, the boy's father, William Ellery Hale, made the elevation of human beings his business. However, where the Rowlands' form of elevation was personal and spiritual, William Hale's business concerned literal and physical elevation. In the aftermath of the great Chicago fire of 1871, the elder Hale's firm installed many of the hydraulic elevators that made new, tall buildings practicable. Over time, his business also expanded to include Europe. Hale's firm even provided the elevators that serviced the new Eiffel Tower in Paris. Thus he elevated people to the greatest heights that mechanical devices could, at that time, attain. Only balloons could take people higher.[4]

George's mother, Mary, remained interested in less tangible concerns. She had "spent her childhood in the shadow of a Calvinistic discipline, dominated by an avenging God, which would have saddened the gayest child." Like Rowland, George Hale early felt the need to acquire instruments and devices, so that his knowledge would continue to grow. He began studies at the Allen Academy at the age of twelve, and was given responsibility for "the 'philosophical instruments'—an air pump, an electric machine, some Leyden jars, a few test tubes and a Bunsen burner." His education also included down-to-earth studies at the Chicago Manual Training School, where the curriculum included "molding and casting, forging and tempering, and machine-shop work, and . . . [Hale] built a steam engine and a 9-inch reflecting telescope."[5]

Although Hale did not meet Rowland until 1889, and never formally studied under him, the Baltimore physicist's analytical expertise appeared in Hale's home laboratory, materialized and embodied in speculum metal. By purchasing a Rowland plane grating, even before he had commenced college studies, the young Hale bought into the enterprise of measuring absorption lines in the solar spectrum. With the instrument, Hale entered the discipline of solar spectroscopy, which remained his particular specialty throughout a long and illustrious career studded with achievements in the organization and promotion of science on national and international scales. The acquisition

---

[3]Wright, *Explorer of the Universe*, 41-42, 43.
[4]On the elevator business, see ibid., 28-29.
[5]Ibid., 30-31, 34, 36.

of light to feed into spectroscopes was always a driving impulse in Hale's career in astronomy; it was a major motive for his creation of large telescopes at Yerkes Observatory, Mount Wilson Observatory, and Mount Palomar Observatory.

As it did in other locales, the grating, and better gratings purchased subsequently, brought with them the mission of the Hopkins School of Light. Through the instrument, the Hopkins school extended its reach into Hale's home laboratory. Through the production and dissemination of gratings, Rowland "taught" spectroscopy to astronomers and physicists far from Baltimore. Use of the instrument was by no means restricted to the Big Spectroscope in the Hopkins physics department, or even to Hale's laboratories. But Hale's deployment, over time, of various Rowland gratings, demonstrated how the production of instruments could spin off new laboratories, a ruled piece of speculum metal at the center of each.

During his college years, Hale continued intermittently to expand his spectroscopic laboratory back home in the Chicago suburbs, when he was not studying both physics and astronomy in Cambridge, Massachusetts. During college, too, Hale took it upon himself to meet Rowland at Johns Hopkins. The meeting at the end of Hale's junior year was memorable, as he told his classmate and lifelong friend Harry Goodwin:

> Rowland was very gruff at first, but before I left—after a visit of over three hours—we were on the best of terms, and I talked . . . freely and easily to him of my plans for future work & c. . . . He showed me all his apparatus, and photos by the dozen, the dividing engines, and everything you could ask. When we went up stairs to look at the grating he even took me into his private office to leave my umbrella & c. There are no flies on Henry A. Rowland, although he does talk in a rather peculiar way. I felt very highly complimented, as he said he was very much pleased with my photos, and that they were about as good as could be made with my grating. I had my grating with me, and he got Schneider to clean it up, and show me how to do it. Rowland also gave me a paper telling all about the practical use of the grating, method of adjustment & c. He said that Brashear has some fine 4" flats, and I can get one of him, as Rowland never sells them himself. He is very much down on Lockyer, and says he is a 'bad man'! This is mostly due to some fuss they had about a grating.[6]

Soon afterwards, Hale was back in Chicago, setting up a grating according to Rowland's instructions.[7]

---

[6]Hale to Goodwin, 6 June 1889, GEHC.
[7]Hale to Goodwin, June 1889 [I], GEHC.

Part of his home became a laboratory: "We have been transforming the 'attic' into an enlarging room, as the space down-stairs was too small to get up to Rowland's scale. We have got the floor laid, and arrangements for light & c. nearly finished. We shall use sky-light as before, and take it up by three mirrors." There was much to do in making an orderly science out of spectroscopy. For example, "in the case of high temperatures, where the metal (for instance) is in the form of vapor, the lines in the spectrum are at present connected by no known law." There was also the continuing need for still better instruments, some of which came from Brashear. Hale wrote, "Brashear is hard at work at my spectroscope and I think he will have it done in time. It is going to be a *beauty*, considerably longer than the one he made for the Lick."[8] (The new Lick Observatory in California then possessed the largest refracting telescope in the world.)

The conversation with Rowland, another with Charles Young at Princeton University (then the College of New Jersey), and some guidance from Edward Pickering at Harvard, led Hale to an invention that formed the basis of his senior thesis at MIT. Called the spectroheliograph, it combined telescope, spectroscope, and photographic camera to select sunlight of only one wavelength and to capture an image of the sun in monochromatic light. The solar image would differ, depending on whether the light photographed came from hydrogen, calcium, iron, or other elements. If one image conveyed too little information, then it was always possible to see the sun in a different light. Although Hale invented the instrument, Brashear actually built it.

A year before graduation, Hale was considering an advanced degree, as he told Goodwin: "Did I tell you that I am thinking of taking a three-year course (Ph.D.) at Johns Hopkins after I return from Europe? Of course I have not decided definitely, but only have it in mind."[9] A year later, Hale told Rowland himself that, while he had considered graduate study in Germany, he was having second thoughts, and might study at Hopkins instead. However, he also informed Rowland that he had a four-inch concave grating, of 10.5-foot focus, ready to use in Chicago.[10] This meant that the Hopkins School of Light had already come to Chicago; Hale did not need to come to the Hopkins School of Light.

During his career, Hale brought about the establishment of four observatories. The best-known were Yerkes (1897), Mount Wilson (founded as a solar observatory in 1904), and Mount Palomar (conceived in 1928 and receiving first light in 1948). The first, however, rose

---

[8]Hale to Goodwin, June 1889 [II], 30 July 1889, 11 August 1889, GEHC.
[9]Hale to Goodwin, 8 July 1889, 14 July 1889, GEHC.
[10]Hale to Rowland, 3 July 1890, HAR.

in the yard of his family home during his college years, and he returned to work in that observatory-cum-laboratory after graduating from MIT. The Kenwood Physical Observatory gave Hale experience in managing the technical and organizational problems of an astrophysical observatory, at a time when astrophysics had little institutional standing (although, contemporaneously, Samuel Langley established the Smithsonian Astrophysical Observatory in 1890). Second, Kenwood later served as a powerful inducement to the University of Chicago, established in 1890, to bring Hale onto the faculty, for when it did so in 1892, Kenwood Observatory became part of the university.

At Kenwood, unsupervised and with a free hand, Hale created the kind of hybrid workplace necessary to the hybrid discipline of astrophysics. Its centerpiece, in place in 1888, was a diffraction grating. When Hale sought, in ensuing years, larger refractor telescopes and then larger reflectors, he always intended to provide more and more light to feed into his spectroscopes, which inevitably attenuated light as they magnified the spectrum. The grating made it possible for Hale to see more of the many spectral lines he encountered in the technical literature, and eventually to join in the debates carried out there. In 1888, Hale confirmed a conclusion by Norman Lockyer that carbon was present in the sun.[11]

In the summer of 1890, Hale ordered a twelve-inch lens for a planned large telescope at Kenwood Physical Observatory, and planned a structure to house the new telescope. He wrote, "We are going to build on the south end of the old K.P.O. The tower will be on the S.W. corner, about 30 ft. in diameter and two stories high. It will be of stone like the house. . . . The telescope will of course be built with special reference to spectroscopic work, and very stiff, so as to carry easily my heavy spectroscope."[12] Kenwood Observatory even had an employee, for "it was found necessary to have a workshop in which a skilled mechanician was almost constantly employed in constructing the numerous pieces of apparatus required in the solar and spectroscopic work."[13] By the summer of 1891, Hale wrote, "The observatory is now practically established as a permanent institution. . . . I shall shortly be regularly elected Director, so as to have a right to that title." It was dedicated in June; the audience included Princeton astronomer Charles Young, John Brashear, and Yale physicist Charles Hastings.[14]

---

[11]Hale to Goodwin, c. August 1888, GEHC.

[12]Hale to Goodwin, 10 August 1890, GEHC.

[13]Hale, "The Yerkes Observatory of the University of Chicago. III. The Instrument and Optical Shops, and the Power House," *ApJ* 5 (1897): 310.

[14]Hale to Goodwin, 9 June 1891, 21 June 1891, GEHC. Wright, *Explorer of the Universe*, 78.

After the completion of the observatory, Hale was back in Europe, meeting with leaders in his field. He traveled to England, Germany, Switzerland, Italy, and France. Some of the acquaintances were competitors as well as colleagues, such as Henri Deslandres. Hale wrote,

> Deslandres fixed up a grating, and using it as I had told him we did, got 7 hydr. lines. He had 2 before, and we have 4, so he is a little ahead. But I am inclined to think that he will have to hustle to keep ahead, if I know myself! Prof. Young has taken up the same work with his new $2600 spectroscope, and has got 3 hyd. lines. Deslandres is first-rate about the thing, and admits that I made the first photos (I was about 3 weeks ahead of him), that I published first, and that our photos are better than his.[15]

Events in Hale's career continued rapidly. In 1892, the University of Chicago appointed him an associate professor of "astral physics" and director of the observatory (his own). Within a few months, Hale began to envision another, larger telescope, utilizing a glass disk that was presently lying abandoned. He told Goodwin, "You may remember that some people in Southern California started a project a few years ago to have the largest telescope in the world. They ordered the glass for a 40 inch, but when the discs were finally finished they found that they had no money to use in completing the instrument. So the unworked discs are for sale." Hale was already looking for the requisite money, in competition with Edward Pickering of Harvard.[16]

The largest operating telescope lens in the world, at thirty-six inches in diameter, was then in the large refractor telescope at Lick Observatory in California. The forty-inch disks represented an opportunity for the University of Chicago to have the largest telescope. With William Harper, the university president, Hale approached Chicago tycoon Charles Yerkes, who had made his fortune in the streetcar business, but who also had a somewhat unsavory reputation that might be improved by a major philanthropic gift. Yerkes agreed to provide the money needed for the telescope and a new observatory to go with it, but only on the condition that the telescope would be the largest in the world, which indeed it would be.[17]

As important as the size of the principal telescope, was Hale's plan that Yerkes Observatory incorporate a physical laboratory along with the observatory. His English friend William Huggins encouraged Hale on this, agreeing that it was best not to devote much attention to traditional astronomical tasks such as determinations of position. Huggins wrote, "I think you do wisely to . . . confine yourself to

---

[15]Hale to Goodwin, 21 August 1891, 8 October 1891, GEHC.
[16]Hale to Goodwin, 17 July 1892, 25 September 1892, GEHC. Wright, *Explorer of the Universe*, 92.
[17]Wright, *Explorer of the Universe*, 95, 97-98.

0astrophysical work; here you have indeed a promised land at your feet, you have only to go up and take it."[18] At the new observatory on Lake Geneva, near Williams Bay, Wisconsin, there would be important provisions for spectroscopic work, including facilities for deploying concave gratings, for photography, for chemistry, and for solar observation.[19] The building, when completed, was a virtual astrophysical cathedral, laid out in the form of a Latin cross.

On the ground floor lay a room devoted to the deployment of Rowland's invention. It was "specially designed to contain a concave grating of twenty-one feet radius, mounted in the ordinary manner." As of 1897, the facility employed gratings of ten-foot and six-foot focus, both brought from Kenwood. For solidity, the mounting for them stood on three brick piers with slate tops. As often with a Rowland grating, the arrangement allowed both sunlight and light from an electric arc to reach the instrument. Light could also reach the grating, through a window, from an adjoining room designated as the physical laboratory: "Thus any desired apparatus can be used in conjunction with the concave grating." Tests showed that the heavy piers used to support various apparatus showed no sign of the tremors that plagued experiments in Baltimore. And careful design made the room light-tight, so that investigators could enter even while photographic plates were being exposed to the spectrum, thanks to double doors. Nearby darkrooms made possible prompt photographic development of the precious spectra captured, and of images recorded elsewhere in the observatory, as well.[20]

Even the purpose of the large telescope was at least partly spectroscopic. Both the great light-gathering power and the large images possible with the great telescope were advantageous for spectrum studies, in that spectra of celestial bodies, as well as conventional photographs of them, would possess better definition by virtue of the greater supply of light. And it was through spectroscopy, in Hale's scheme, that astronomy could be transformed into astrophysics. As Hale recounted, "Although important astrophysical observatories existed at that time in France and Germany, and provision had been made for astrophysical work in the national observatories of England and Russia, there was no observatory in North America which was completely equipped for such investigations."[21]

---

[18]Huggins to Hale, 10 December 1892, George Ellery Hale Papers, Microfilm Edition (GEHM), ed. Daniel J. Kevles (Washington, DC: Carnegie Institute of Washington, 1967).

[19]Wright, *Explorer of the Universe*, 108-109.

[20]Hale, "The Yerkes Observatory of the University of Chicago. II. The Building and Minor Instruments," *ApJ* 5 (1897): 264-266.

[21]Ibid., 254-255.

Allegheny Observatory in Pittsburgh "probably represented more nearly than any other such a union of astronomy and physics as an astrophysical observatory implies," but lack of large instruments and sufficient staff hobbled efforts there. In Hale's view, Harvard and Lick Observatories lacked adequate physical laboratories. And the School of Light at Johns Hopkins was "unable, for lack of instrumental means, to carry their fundamentally important researches beyond the artificial boundaries of physics into the realm of astronomy." Although Samuel Langley carried out research at the Smithsonian, Hale felt that Congress "had not seen fit to provide funds for establishing on a proper scale a national astrophysical observatory."[22]

Hale concluded, "In short, in spite of the wealth of observatories and laboratories in the United States, and the numerous contributions to astrophysics of Rutherford [sic], Draper, Young, Pickering, Langley, Rowland, Keeler, Michelson, Hastings, and many others, there was yet to be founded an observatory which should adequately represent both the astronomical and physical sides of astrophysical work."[23] Thus, to succeed where others had failed, Hale appealed to the need of businessman Charles Yerkes, with his tainted public image, for a chance to redeem his unsavory reputation. Although, as plans proceeded and incidental expenses arose (Yerkes did not always give money as readily as Hale felt he had promised to do) the finished observatory reflected Hale's plans for astrophysics very well.[24]

Never one to pursue a single project at a time, Hale had already begun to give his elected discipline formal recognition by launching a regular publication forum, and he arranged for Henry Crew, Joseph Ames, and James Keeler, all alumni of the Hopkins school, to be his three fellow editors of an "astro-physics" department of the journal *Astronomy and Astrophysics*, which first appeared in January, 1892. The

---

[22]Hale, "The Yerkes Observatory II," 254-255. Langley became Secretary of the Smithsonian Institution in 1887, and began building an astrophysical observatory there in 1889, a project partially subsidized with $5,000 from Alexander Graham Bell. He had wanted a better site " 'away from the tremors of the soil' and busy city streets" surrounding the Smithsonian, but failed to place the observatory in a zoological park as he had hoped. In 1891, Langley secured an appropriation from Congress for maintenance of the observatory. Working conditions were unfavorable, however, as temperatures inside, during the summer, reached 100 to 110 degrees Fahrenheit. Bessie Zaban Jones, *Lighthouse of the Skies: The Smithsonian Astrophysics Observatory, Background and History, 1846-1955* (Washington, DC: Smithsonian Institution, 1965), 112, 120, 127, 133.

[23]Hale, "The Yerkes Observatory II," 255. Henry Draper, an amateur, was one of the first to photograph the spectrum of a star, in 1872. His father, John William Draper, may have been the first to photograph the diffraction spectrum of the sun, in 1844. Charles A. Young was an astronomer at Princeton University. Charles Hastings, a Yale physicist and an authority on optics, was for a while a colleague of Rowland at Johns Hopkins. On Keeler, *vide infra*.

[24]On friction between Hale and Yerkes over funding of the project, see again Wright, *Explorer of the Universe*, 97-102. Some of the incidental expenses included instruments supplied by the indefatigable Brashear.

other department was called "general astronomy," and the editor of that section was W.W. Payne of Minnesota, who allowed the new journal to supersede his earlier *Sidereal Messenger*. The scope of Hale's section was mostly confined "to the more technical subjects connected for the most part with spectroscopic work."[25] His own published research had only started to appear in earnest in 1891, when he published twelve articles, of which four appeared in foreign journals. The majority concerned the sun, a personal research emphasis that continued throughout his career. In 1892, when *Astronomy and Astrophysics* began to appear, Hale increased his output. Of the twenty-six articles that he published that year, all but four appeared in *Astronomy and Astrophysics*.[26]

Already in 1892, however, Hale's aims for putting astrophysics into print, as well as reifying it in observatories, kept growing larger. Between 1892 and 1895, the astrophysics section of the joint journal gave way to Hale's *Astrophysical Journal*, published in Chicago. The journal gave definition and a common agenda to a discipline that had previously lacked them. In the briefest terms, the scope of the journal included "all investigations of radiant energy, whether conducted in the observatory or in the laboratory." The journal was intended especially for spectroscopists and other investigators who would otherwise be uncertain whether to publish in a physics journal or an astronomical journal. Wrote Hale, "The astronomer and physicist should be able to meet on common ground." In addition to joining together these previously separate disciplines, the journal also joined together scientists of different nations, especially drawing in Europeans. Hale was very active in recruiting editors in Europe, "for it was felt from the first that unless the journal were made truly international in character it could not be a success." To represent their respective nations, five associate editors were named: M.A. Cornu of France, N.C. Duner of Sweden, William Huggins of Britain, P. Tacchini of Italy, and H.C. Vogel of Germany. The board also included five associate editors prominent in the U.S.: Hastings of Yale, Michelson of Chicago, Pickering of Harvard, Rowland, and Young of Princeton. In addition, Hale kept the Hopkins-derived nucleus he had formed for the previous journal: Keeler became co-editor with Hale, and Crew and Ames became assistant editors. Rowland and Young, who had earlier directed Hale toward his undergraduate thesis on solar astronomy, lobbied for the word "spectroscopy," and so the subtitle of the journal read: "An International Review of Spectroscopy and Astronomical Physics."[27] Because of the diverse nationalities

---

[25]Hale, "The Astrophysical Journal," *ApJ* 1 (1895): 80.
[26]Walter S. Adams, "George Ellery Hale, 1868-1938," *BMNAS*, 21 (1941): 219-221.

represented on the editorial board, William Huggins told Hale, even before the first issue appeared, "You deserve international thanks."[28]

From the first, the work published showed great diversity. In the initial year, almost one-quarter of the articles concerned the moon and planets; twenty-one percent concerned stellar observation or stellar spectroscopy; seventeen percent concerned instruments; fourteen percent were about solar observations; twelve percent addressed laboratory spectroscopy; and six percent each touched on physical theory or studies of the nebulae and the distribution of stars. Over the first five years, stellar observation and spectroscopy, and instruments, each continued to represent about twenty percent of the major articles. A further twenty-one percent concerned laboratory spectroscopy, sometimes in conjunction with astronomical spectroscopy. Observations of the sun accounted for ten percent, and observations of other parts of the solar system represented twelve percent. Five per cent concerned physical theory, and six percent concerned astronomical theory. The remainder covered either stellar distribution, nebulae, or the issue of standards.[29] These statistics do not include Rowland's tables of some 20,000 solar absorption-line wavelengths, which appeared in installments during the first years of publication. Hale had lobbied Rowland since 1891 for permission to publish his tables; at that earlier date, the list only ran to 1,000 absorption lines.[30] (The demand came from other sources, as well. In 1893, Huggins had asked Hale to send a list of Rowland's standard lines.)[31]

Hale also made a point of contributing to the *Astrophysical Journal* himself. In 1895, the first year of publication, Hale contributed fifteen papers. The following year he contributed seven.[32] One of the papers expanded on insights attained at the Hopkins school by Jewell, Humphreys, and Mohler. The work of these three suggested that variations in the exact position and intensity of spectral lines might provide information on variation of the conditions in the solar or laboratory matter that gave rise to the spectral lines. Until recently, Hale wrote, "The Fraunhofer lines have been regarded as fixed marks of reference, subject to no possible change in position. Moreover, but

[27]"Notice," *ApJ* 6 (1897): facing p. 271. "The Astrophysical Journal," *ApJ* 1 (1895): 81-82. Wright, *DSB*, 28. Osterbrock, *Keeler*, 206. Hale had been promoting the idea of a specialized astrophysical journal for years, and by 1891 had already received endorsements from Huggins, Vogel, Cornu, Langley, Young, Pickering, Keeler, and numerous others. Hale to Rowland, 21 November 1891, Letterpress Book #1, GEHC.

[28]William Huggins to Hale, 11 December 1894, GEHM.

[29]Article distribution quantified from a survey of *ApJ*, 1895-1899.

[30]Hale to Rowland, 21 November 1891, Letterpress Book #1, GEHC.

[31]Huggins to Hale, 6 May 1893, GEHM.

[32]Adams, "Hale," 223-224.

little evidence has been advanced to show that any of the Fraunhofer lines vary in intensity, though that a secular change is going on is indicated by the variety in type of stellar spectra." Variations in line intensity and position in the solar spectrum potentially held clues to the temperature, pressure, and motion of gases in the sun. Hale asserted, "Every photograph of the solar spectrum taken with high dispersion must now be regarded as a document of great value, which may ultimately reveal irregular or periodic changes in the condition of the gases and vapors of the solar atmosphere."[33]

Although Hale was an important contributor, the purpose of the journal was to promote the discipline, and not necessarily Hale himself. The journal competed favorably with other fledgling efforts such as *The Physical Review*, founded in 1893 at Cornell. Kevles writes, "Lacking prestige, the *Review* lost much of the good work of the younger men to foreign journals or, in the case of spectral studies, to the newly founded *Astrophysical Journal* in the United States."[34] Physicists contributing to Hale's journal saw their work appear in distinguished company. Rowland was, after all, a leader of the first rank in the physics community, and the appearance of his definitive tables lent authority to the hybrid journal. It also demonstrated that the implications of Rowland's own research, as well as his instrument-making, crept beyond the boundaries of physics proper. Both together provided a firmer basis for the new discipline that Hale was promoting.

Kenwood, Yerkes, and the *Astrophysical Journal* were still not enough for Hale. In fact, Hale never stopped planning larger and better observatories, and until old age he never stopped promoting scientific associations on ever-larger scales. In the new century he established a new home for his various enterprises by starting an observatory on Mount Wilson in California, within a decade after the completion of Yerkes. As had been the case at Kenwood, Hale initiated the project entirely on his own. In late 1903, he left Chicago for California, ostensibly under the auspices of the Yerkes Observatory. "But, at the moment, there was no financial support for such an expedition, except what was coming out of his own pocket."[35] He had already investigated the mountain, in summer, with W.W. Campbell of Lick Observatory, on behalf of the Carnegie Institution of Washington, to consider situating an observatory there. Hale had returned to Chicago with a favorable opinion of the location, but faced inaction on the part of the institution and of Andrew Carnegie himself. However,

---

[33]Hale, "Note on the Application of Messrs. Jewell, Humphreys and Mohler's Results to Certain Problems of Astrophysics," *ApJ* 3 (1896): 156-157, 159.

[34]Kevles, *The Physicists*, 81.

[35]Wright, *Explorer of the Universe*, 172.

even before receiving the negative opinion, Hale made independent plans to keep his idea alive: "unable to wait, he decided to go ahead to California and gamble on his own."[36]

He returned to assess further the conditions on Mount Wilson. On this second trip, after reaching the top, he climbed a further sixty-eight feet, up a pine tree, with a small telescope, to test the seeing conditions well above ground and to avoid potential air turbulence that might have skewed his observations at ground level. The lack of funding did not deter him in his climb. He wrote to Charles Mendenhall: "It is true that no appropriation was made this year for the Solar Observatory scheme. There is good reason to hope for favorable action next year, however, and at any event I am going on with my study of the conditions for solar work in the mountains here."[37]

In the long run, the difference between ground level and tree level was significant to Hale. After the observatory gained more formal recognition, it built two solar tower telescopes, reaching sixty and 150 feet above ground (completed in 1908 and 1912, respectively). Limited funding had come from the Carnegie Institution toward the end of 1904,[38] and the facility was named the Mount Wilson Solar Observatory. Even before securing that support, however, Hale began to render the peak habitable, initially by renovating an existing building at the summit early in 1904. This "required funds, most of which continued to come out of Hale's own pocket."[39]

Thus the future site of sixty-inch (1909) and 100-inch (1917) reflecting telescopes, with which astronomers later changed cosmology profoundly, started, not unlike Kenwood Observatory, as a base from which Hale could, on his own resources, use Rowland's instrument to take high-dispersion spectra of the sun, which was for Hale a typical star. Hale's ambitions were so clear to him that, in 1906, when the observatory was barely established, he turned down both the presidency of the Massachusetts Institute of Technology and the office of secretary of the Smithsonian Institution, in order to continue building a material basis for a new way of seeing the heavens. The goal of securing instruments that would gather more light was paramount because, Hale wrote, "astronomy subsists on light."[40]

---

[36]Ibid., 171.

[37]Ibid., 174-175. Hale to Mendenhall, 13 January 1904, GEHM.

[38]Robert S. Woodward to Hale, 20 December 1904, GEHM.

[39]Wright, *Explorer of the Universe*, 178.

[40]Hale, *The Study of Stellar Evolution* (Chicago: University of Chicago Press, 1908), 13. On reasons for declining the Smithsonian, Hale wrote to Simon Newcomb, "I confess that I have thought many times of the interesting problems in store for the next Secretary, but the work here must be carried through as planned, and the pleasure of doing this will be so great that other positions could not be as attractive as the one I now hold." Hale to Newcomb, 19 March 1906, GEHM.

The tenacious independence of Hale's approach, including his willingness to stake personal funds at Kenwood and again at Mount Wilson, were important to the emergence of astrophysics. The new field, during much of the nineteenth century, received important impetus from amateurs and outsiders to the profession of astronomy. (Significant exceptions in America before 1890 were Pickering at Harvard, Young at Princeton, and Langley. Solar physics was the first branch of astrophysics to become professionalized.) In England, J. Norman Lockyer began work in astrophysics as an amateur, and subsequently became director of the Observatory of Solar Physics and founder of the journal *Nature*. In the same country, Huggins pioneered stellar spectroscopy in a private observatory that he had constructed for himself.

John Lankford has elucidated the role of amateurs in the latter half of the nineteenth century:

> In England and America the new specialty of astrophysics did not develop in the university context. Astrophysical research often started outside established observatories. Further, the problems astrophysicists initially sought to examine placed a premium on the design, construction, and manipulation of complicated research equipment, rather than on theoretical knowledge. Here was a field in which amateurs might excel.[41]

Astrophysics lacked a strong institutional base until the 1890s. Before the late nineteenth century, spectroscopy was not one of the routine tasks of an observatory; it was not practiced by professionals. Thus astrophysics was practiced for the most part outside the 143 observatories that existed in the U.S. in 1882.[42] Institutional and financial support were gradually forthcoming, however, and astrophysics became more accepted toward 1910.

Martin Harwit has asserted that discoveries of new phenomena in astronomy often come from researchers who are in some sense outsiders.[43] The rise of astrophysics, with the phenomena which it brought within the purview of astronomy (essentially spectra), may be an instantiation of this rule. Two of the principal founders of astrophysics in the U.S.—Hale and his co-editor at the *Astrophysical Journal*, James Keeler—received a significant portion of their training outside of astronomy. Hale's official major at MIT was physics, not astronomy.

---

[41]John Lankford, "Amateurs and Astrophysics: A Neglected Aspect in the Development of a Scientific Specialty," *Social Studies of Science*, 11 (1981): 277.

[42]Number from Nathan Reingold, *Science, American Style* (New Brunswick: Rutgers University Press, 1991), 65.

[43]Martin Harwit, *Cosmic Discovery: The Search, Scope, and Heritage of Astronomy* (New York: Basic Books, 1981), 20.

Astrophysics necessarily involved the importation of new instruments and techniques. It benefited from the research program that Rowland the physicist fostered. In the time of Hale and Keeler, "the new word 'astrophysics' meant the use of spectroscopes and spectrographs on telescopes to analyze the light from celestial objects that reaches the surface of the earth."[44] Although Rowland is always described as a physicist and never as an astrophysicist, solar physics has been described as the "mother of astrophysics."[45] Thus the institutional support that Rowland enjoyed at Johns Hopkins when he embarked on his monumental study of the solar spectrum was a vital resource. The new discipline faced an uphill struggle.[46]

The creation or acquisition of adequate instruments was an essential part of the struggle. The discipline required observational engineering as a prerequisite to new ways of seeing. Thus it may be intelligible why, in 1920, when the eminent Dutch astronomer J.C. Kapteyn looked back at Hale's early career, he concluded: "I really begin to think that the true recipe for making a first rate astronomer is: Take an engineer and teach him some astronomy, not: take an astronomer and teach him some engineering."[47] It may have taken an engineer's training to devise various means, as Hale and co-workers did, by which, in Wright's words, "a star could be brought into a constant-temperature physical laboratory."[48] Of course, aside from his course in physics at MIT, Hale's informal education in astronomy was substantial. While at MIT, he worked part-time at Harvard College Observatory. He also looked outside universities completely: "In his spare time he read and abstracted everything he could find on astronomy and spectroscopy at the Boston Public Library."[49] Although Hale's postsecondary education at MIT was practical in nature, he started as an avocational astronomer, and in this respect Hale almost personified the transition of astrophysics from an amateur to a professional science.

The characterization of Hale and his colleague Keeler as outsiders, in any sense, to astronomy, applies only to their early careers. Although both studied astronomy to some extent during their formal education, it was not their primary emphasis. Both brought to astronomy methods that they had learned on their own initiative. Keeler, Hale, and another innovator, John Anderson, were primarily educated in physics. Beneficiaries of the Hopkins school thus appeared as

---

[44]Osterbrock, *James E. Keeler*, 1.

[45]Eugene N. Parker, quoted in Hufbauer, *Exploring the Sun*, 193.

[46]Robert W. Smith, *The Expanding Universe: Astronomy's 'Great Debate' 1900-1932* (Cambridge: Cambridge University Press, 1982), 6.

[47]Kapteyn to Hale, 23 October 1920, quoted in Wright, *Explorer of the Universe*, 143.

[48]Wright, *Explorer of the Universe*, 150.

[49]Wright, "Hale," *DSB*, 27-28.

important enablers, significant not only for their interpretation of natural phenomena, but more importantly for their role in making the precise observation of new phenomena possible.

By the time Hale graduated from MIT in 1890, the older Keeler was already a professional astronomer. As Hale's biographer reported: "More than anyone else in the country, Keeler shared Hale's interest and belief in the embryonic science of astrophysics. There was no one Hale was more anxious to meet."[50] As often in life, one important influence on Keeler in his youth had been his father, "from whom he inherited his extraordinary taste for mechanical pursuits, and to whom he also owed his early instruction in the use of tools and the design and construction of innumerable conveniences of daily life."[51]

In 1875, when Keeler was eighteen, he purchased two lenses and used them to make a telescope. From his observations, he made color drawings of the planets. Two years later, he constructed "a meridian instrument, which he made from a marine spyglass with the tools at command in his own home."[52] Keeler's other homemade instruments included a quadrant and a chronometer.[53] Keeler first considered career paths in astronomy that did not involve college education. One possibility was a position in Cambridgeport, Massachusetts, in the optical instrument works of Alvan Clark, who made many of the most important instruments, especially telescope lenses, in American astronomy. However, there were no positions available. Instead, Keeler began to think about further education, and his subsequent affiliation with the new university in Baltimore continued to profit him well after he received the B.A. in 1881. While Keeler was at Hopkins, a wealthy amateur astronomer, Charles Rockwell, helped to pay for his education.[54]

Although Keeler matriculated as an undergraduate at Johns Hopkins in 1877, the second year of operation of the university, he was only indirectly a student of Rowland, because the latter did not offer undergraduate courses. However, Rowland may have had a greater influence on undergraduates during the first years of the university than he did later, because it was only in 1884 that President Gilman, concerned about Rowland's shortcomings as a teacher, finally asked him to sign a formal statement to the effect that he would not interfere with undergraduate physics instruction as offered then by Charles Hastings, Keeler's primary mentor in physics.[55]

---

[50]Wright, *Explorer of the Universe*, 71.
[51]Charles S. Hastings, "James Edward Keeler, 1857-1900," *BMNAS*, 5 (1905): 233.
[52]Ibid., 234.
[53]Sally H. Dieke, "Keeler, James Edward," *DSB*, Vol. 7 (1973), 270.
[54]Osterbrock, *Keeler*, 8-9, 14, 165.
[55]Miller, dissertation, 290. Rowland to Gilman, 12 March 1884, DCG.

There was no regular astronomy professor at Hopkins during Keeler's college years, but the eminent astronomer Simon Newcomb, who worked at the Nautical Almanac in Washington, D.C., came frequently to lecture, and taught Keeler the astronomy of the time. (Later, in the 1890s, Keeler recognized a tension between astrophysics and the older discipline of positional astronomy, as practiced by Newcomb.) At Hopkins, Keeler "made many close friends among his classmates and the graduate students." He majored in physics and German, and subsequently pursued post-baccalaureate studies in Berlin, where his undergraduate work was well respected.[56]

During Keeler's final undergraduate year, astrophysicist Samuel Langley visited and lectured. After graduation, Keeler became an observing assistant to Langley at Allegheny Observatory near Pittsburgh. Keeler's first challenge was to become familiar with Langley's instruments—telescopes, mirrors, gratings, prisms, bolometers, thermometers.[57] This practical experience, essential to astrophysics, was only available on the job, and not in any formal training program. Langley himself was largely self-taught in astronomy, and had only recently invented the bolometer (a thermometer used to detect thermal radiation and to study infrared spectra), which Brashear built for him.[58]

During this employment, Keeler got to know Brashear, who in fact made almost all of Langley's instruments. (Langley had "discovered" Brashear, arranging philanthropic support for him after recognizing his instrument-making ability.) For a while, Keeler also worked in the laboratory of Hermann von Helmholtz in Berlin. He gained new ideas, but he also brought instruments with him: a bolometer lent by Langley and a grating given by Rowland. His observational work in Germany was mostly limited to the physical laboratory rather than the observatory, but he surely kept astronomy in mind. He studied the absorption of infrared radiation by carbon dioxide contained in a tube, and was able to show that the amount of absorption was significant. Thus it followed that part of the radiant energy in sunlight must be absorbed by the atmosphere before it could be measured by bolometers.[59]

In general, absorption lines that might be caused by carbon dioxide in the terrestrial atmosphere or gases in the outer atmosphere of the sun were more common than emission lines. (The 20,000 solar lines tabulated by Rowland were all absorption lines.) By studying spectral lines, astrophysicists thus frequently studied the matter that lay between them and the ultimate sources of light, such as the deeper

---

[56]Ibid., 9-10, 18, 248.
[57]Ibid., 20.
[58]Ibid., 15. Gaul and Eiseman, *John Alfred Brashear*, 78.
[59]Osterbrock, *Keeler*, 27, 33.

layers of the sun, which often gave off continuous spectra, which were subsequently marked by absorption lines due to overlying matter. For the most part, absorption caused by the terrestrial atmosphere was merely an obstacle to be identified and eliminated as an object of interest, a task also taken up by William Meggers (Johns Hopkins Ph.D., 1917) in the following century.

While in Germany, in addition to his experiments in Helmholtz's laboratory, Keeler also attended lectures by Heinrich Kayser, further expanding his grasp of laboratory spectroscopy. It was also in Berlin that Keeler met Henry Crew, where they both studied under Helmholtz and Kayser, and there Keeler directed him toward graduate training at Hopkins.[60] Upon his own return to the U.S., Keeler resumed work at Allegheny, where he remained until 1886. Then, from 1886 to 1891, he worked at Lick Observatory, where he was the first professional astronomer; among other duties, he operated a time service for commercial interests.[61] In 1891, he returned to Allegheny as director, and remained there until 1898, when he became the director of Lick. Like Rowland, Michelson, and Hale, Keeler sometimes turned to Brashear for material assistance. The latter supplied a spectroscope for Lick in about 1888, and Keeler used it to study the radial velocities of nebulae. After Keeler's return to Allegheny, Brashear supplied another spectroscope, which Keeler used to study the radial velocities of nebulae, stars, and planets. This second device included both a Rowland plane grating and a train of three prisms.[62]

During his varied career, and over the distances, Keeler continued to receive materials and expertise from Johns Hopkins. In 1889, he studied Doppler shifts in stellar spectra, and he was able to use a Rowland grating to produce spectra of the brightest stars that he observed. He compared them to spectra from a laboratory light source— in this case the flame, colored by salt, of an oil lamp. Hearnshaw writes, "Keeler's work was the first successful observation of grating spectra from a celestial source other than the sun. His success can be attributed to his use of a high quality Rowland grating, to his use of a large telescope, and to his restriction of observations either to emission-line nebulae or to very bright stars."[63]

Keeler also used a Rowland grating to measure two green emission lines from a nebula in the constellation Orion. The two well-known lines appeared in many nebulae; some observers thought they represented an element ("nebulium") that did not exist on Earth.

---

[60]Crew to Robert S. Woodbury, 4 December 1935, HCM. Osterbrock, *Keeler*, 32.
[61]Dieke, "Keeler," 270.
[62]*John A. Brashear*, 245.
[63]Hearnshaw, *The Analysis of Starlight*, 10.

Keeler chose a laboratory line from lead as a reference line for comparative measurement, and asked Rowland to tell the exact wavelength of the reference line. Rowland's researches, however, had not yet extended to lead.[64] Until recently, Keeler "had never actually taken a photographic spectrogram"; he had only observed spectra directly at the telescope. As part of the remedy, he directed Brashear to build a spectrograph, and in the meantime turned to Baltimore for more background. "He quizzed Henry A. Rowland's assistant, Lewis Jewell, to find out all the photographic methods used in the spectroscopic laboratory at Johns Hopkins." Keeler then visited Baltimore and talked directly with Rowland about principles of photography. In 1891 Keeler wrote a paper on the nebular lines; while the article awaited publication, Jewell finally measured the lead comparison line. Keeler then changed all of his wavelengths to conform to the Hopkins standards. In 1893, Keeler needed to learn about photographic materials, and turned to an outlying alumnus of the Hopkins school, namely Crew, at Northwestern University near Chicago.[65] (Crew had spent the year 1891-1892 at Lick Observatory, studying Doppler shifts in stellar spectra and attempting to use concave gratings to photograph stellar spectra.)[66]

In addition to know-how, Keeler also looked to Hopkins for help in maintaining his own standing in the scientific community. As an alumnus, Keeler regularly sent letters and reprints to President Gilman, to promote awareness of his own work. In 1897, when Gilman organized a series of lectures on "Recent Advances in Astronomical and Physical Sciences," he included Keeler among the speakers. Keeler also turned to Rowland for endorsement. During his study of the enigmatic nebular lines, Keeler wanted "the acknowledged leader of American spectrocopy as his advocate." Rowland, for his part, "was deeply interested in the nebular problem." Prior to publication, Keeler shared his results with Rowland, as well as Huggins in England.[67]

In 1893, Keeler gave a formal presentation of his work on the nebular lines at the Chicago Columbian Exposition. He showed that the lines did not coincide with any known solar or laboratory lines, and were not due to magnesium, as Norman Lockyer had suggested. The paper "undoubtedly captured the interest of Rowland, the great physicist who had been too busy to pay any attention to Keeler in his student days at Johns Hopkins." Keeler had successfully enlisted both the technical expertise and the tacit approval of his alma mater; also,

---

[64]Osterbrock, *Keeler*, 94.
[65]Osterbrock, *Keeler*, 126-127, 133, 137.
[66]William F. Meggers, "Henry Crew, 1859-1953," *BMNAS*, 37 (1964): 38.
[67]Osterbrock, *Keeler*, 94, 106, 221.

he "had arrived as a leader in his field." Yet he had not produced an explanation of the nebular lines. Rather, he presented a definitive observation, a wavelength measurement, of a novel and mysterious phenomenon.[68]

On the exact wavelengths, Keeler had found in 1890 that his measurements differed from those of Huggins. He initially doubted his own results, but when he found a comparable discrepancy between his measurements of the spectrum of Arcturus and Huggins's measurements of the same light, he turned to the measurements of Vogel, which confirmed Keeler's results rather than those of Huggins. "Then Keeler had evidence that all his tests had not lied, that his nebular wavelength measurements were also doubtless correct, and that Huggins was simply wrong, almost certainly because of his inadequate instruments. Keeler was quietly exultant."[69] Thus, Keeler's work gained international stature.

The enigma of the nebular lines persisted for more than thirty years, during which time astronomers generally attributed them to nebulium, even though the putative element had never been found in the laboratory. The man who eventually solved the mystery, American physicist Ira Bowen, deduced in 1928 that the lines must come from doubly ionized oxygen, at a density too low to reproduce in the laboratory. (Later, in 1946, Bowen became director of the Mount Wilson and Palomar Observatories.)

Although measurement of the nebular lines may have established Keeler's name, Hale's friend and colleague also reached important findings concerning Saturn. With the relatively new tool of photography, Keeler was able to record spectra from Saturn's rings in 1895. After measurements of Doppler shifts in the spectra, he concluded that the rings did not rotate as solid objects, but rather revolved more quickly in sections closer to the planet, and more slowly in outer sections, as rings composed of myriad separate objects, or particles, would do. He thus obtained observational evidence confirming earlier theoretical work by James Clerk Maxwell, who had deduced mathematically in 1857 that the rings were particulate in composition. (Studies of Saturn likely captured the imagination of a number of prominent physicists. After 1900, some at the Cavendish Laboratory at Cambridge University in England envisioned models of the atom that were reminiscent of the sixth planet and the surrounding rings. Henry Crew dubbed this model "the Cambridge atom," or alternatively "the Saturnian atom." This atom contained a large number of electrons "distributed throughout the same orbit." The differing

---

[68]Ibid., 94, 99, 106, 196, 198, 221.
[69]Ibid., 94-97.

properties of different elements could be attributed to differing configurations of rings: "then one element differs from another mainly in
the number, disposition and character of the rings which surround the
central attracting charge.")[70]

Keeler was also quick to list milestones achieved by others in astronomical spectroscopy. One was the observational proof of the occurrence of the Doppler effect, by cross-comparison between
spectroscopic results and those achieved by the older astronomy of
position. Other milestones were: the discovery of helium; extended
observations of the series in the hydrogen spectrum that was later
named for Balmer; solar motion; the discovery of numerous spectral
series in the laboratory; and the dynamic behavior of spectroscopic bi-
nary stars.[71]

However, even taking these discoveries into account, Keeler allowed that the primary motivation for practicing astrophysics could
not be pragmatic: "With respect to practical usefulness, . . . astrophysics does not possess the same claims to consideration as astronomy." The latter was applicable to time signals, as at Lick, and to
navigation, surveying, and other uses. The aims of astrophysics were,
in Keeler's day, less tangible: to discover "unknown laws and principles," to "throw light on the dark places in nature," and to "elevate
the thoughts and ennoble the minds of men."

Although Keeler was not a popularizer, he held hope that findings from astrophysics might attract wide public attention:

> A star, regarded as a center of attraction, or as reference point from
> which to measure celestial motions, awakens little enthusiasm in
> the popular mind; but a star regarded as a sun, pouring out floods
> of light and heat as a consequence of its own contraction, torn by
> conflicting currents and fiery eruptions, shrouded in absorbing
> vapors or perhaps in vast masses of flame, appeals at once to the
> popular imagination.[72]

Keeler also had higher principles in mind. He admitted that the
apparatus of astrophysics might lack aesthetic appeal: "There may be
some who view with disfavor the array of chemical, physical, and electrical appliances crowded around the modern telescope, and look back
to the observatory of the past as to a classic temple." But he believed
that the clutter of apparatus assisted his search for an uncluttered
"unity of nature" undivided by disciplinary boundaries. The new
study of astrophysics would promote "an exchange of knowledge"

---

[70]Crew, "Fact and Theory in Spectroscopy," *Science*, 25 (1907): 4-5.
[71]James E. Keeler, "The Importance of Astrophysical Research and the Relation of Astrophysics to Other Physical Sciences," *ApJ*, 6 (1897): 279-282.
[72]Ibid., 275.

between neighboring disciplines that would tend "to bring them into more perfect coordination." In this view, Keeler shared a vision with Hale, who "looked back with nostalgia to a simpler era characterized by a greater degree of unity."[73] The two differed mainly in the social context of their science: Keeler was appointed to existing institutions, whereas Hale created new institutions.

In each of the observatories that Hale created, from Kenwood to Mount Wilson, Hopkins gratings were essential. Although big telescopes were the feature that attracted the most attention, spectroscopic instruments were equally important to the new discipline. And Hale was not satisfied simply to own gratings from Johns Hopkins, or to collaborate with Keeler and other Hopkins alumni. After he began to put Mount Wilson Observatory into operation, Hale wanted the mechanical expertise that lay at the base of the Hopkins School of Light. As soon as he started to establish the observatory, Hale sought assurances from the trustworthy Brashear that there would be a ready supply of Hopkins-made instruments: "I trust that Mr. Jewell will succeed in getting a perfect grating very soon."[74] Hale constantly felt the need for more and better from Hopkins. The best grating available for the sixty-foot solar tower telescope, upon its completion, Hale reported, was "the 4-inch plane that I used in my work at Kenwood." A five-inch plane grating, on loan from Hopkins, was then "doing yeoman service in the laboratory." During the early days at Mount Wilson, Hale made his most important discovery with the same four-inch grating that he had acquired for the Kenwood Observatory in 1889. He found that lines in the spectrum of sunspots showed a broadening, due to the Zeeman effect, suggesting the presence of strong magnetic fields. But he made this discovery [75]

In July 1904 Hale expressed impatience. He asked Joseph Ames about Jewell's progress in grating production, and informed Ames that he hoped to secure funds for a ruling engine at Mount Wilson. Excellent observing conditions there sharpened the need for good instruments: "The sky remains perfectly clear day after day."[76] Ames replied that he

---

[73]Ibid., 271, 273, 277, 288. Nathan Reingold and Ida H. Reingold, *Science in America: A Documentary History 1900-1939* (Chicago: University of Chicago Press, 1981), 194.

[74]Hale to Brashear, 26 February 1904, GEHM. Jewell operated the ruling engines after Rowland and Schneider.

[75]Hale to Ames, 24 January 1908, 16 November 1909, GEHM. Hale also subsequently pursued the question of whether magnetism existed in the sun beyond the boundaries of sunspots. Thus, he "led the project to measure a general solar magnetic field." But Norriss Hetherington suggests that, soon, "Hale's anxiety to find the expected result had overcome his scientific judgment." Later measurements of the solar magnetic field differed greatly from the values found by Hale's observers. Hetherington, *Science and Objectivity* (Ames: Iowa State University Press, 1988), 73, 79-80.

[76]Hale to Ames, 21 July 1904, GEHM.

would be glad to help Hale arrange for his own ruling engine,[77] but little changed over the next few years. In 1905, Mount Wilson possessed a grating with a focal length of thirteen feet. With it, astronomers photographed the spectrum of the star Arcturus over three successive nights (the faintness of starlight requiring the cumulative exposure), during which the temperature of the spectrograph was kept constant to within 1/20 degree.[78] The difficulty of making such an observation kept Hale occupied with "the perennial problem of gratings." He wrote to Ames: "I beseech you, in the name of the Prophet, to do something for us in the matter of gratings." He so beset Rowland's successor about his needs that Ames suggested that Hale employ Jewell himself.[79] Hale apparently did not take up this offer, but the idea of operating ruling engines at Mount Wilson took root. By 1905, he began to pursue it.[80]

Hale was eventually able to hire John A. Anderson (Johns Hopkins Ph.D., 1907), an associate professor at Hopkins who learned spectroscopy under Ames, and taught astronomy there. Upon his appointment as an instructor at Hopkins in 1908, Anderson was given charge of the ruling engines. He thoroughly rebuilt one of them, "and then ruled a substantial number of gratings with higher resolving power, less scattered light, and weaker 'ghost' intensities than any produced before."[81]

Hale informed Ames that he ultimately hoped to see a grating with a ruled surface of eight inches by ten inches. Ames replied that Anderson's goal was a surface of five inches by nine inches. Hearing of a man with aspirations similar to his own, Hale soon sought Anderson's presence at Mount Wilson. As he told Ames, "We are desperately in need of large gratings." In order to reduce his outside dependency, Hale said, "it does appear that a ruling machine should be as much a part of our equipment as a lathe in the shop."[82]

Hale considered astrophysics to be an "applied science,"[83] but he probably had in mind the application of skills such as Anderson's to problems of celestial observation, rather than the application of astrophysical knowledge to the problems of life. Hale did not need

---

[77]Ames to Hale, 17 August 1904, GEHM.

[78]Hale to Charles A. Young, 24 July 1905, GEHM.

[79]Hale to Ames, 24 January 1908; Ames to Hale, 20 October 1909, GEHM.

[80]Hale to Charles Mendenhall, 3 March 1905, GEHM. Hale also told Woodward, his patron at the Carnegie Institution of Washington, "I know of nothing more important for the development of astrophysical research than large and perfect gratings. A single grating with a ruled space 10 X 15 inches would easily be worth $2,000 to us, and probably a good deal more." Hale to Woodward, 30 April 1906, ibid.

[81]Ira S. Bowen, "John August Anderson, 1876-1959," *BMNAS*, 36 (1962): 1-3.

[82]Hale to Ames, 14 June 1910; Ames to Hale, 24 June 1910; Hale to Ames, 27 November 1911, GEHM.

[83]Hale to Ernest Rutherford, 1 June 1914, GEHM.

Anderson to articulate the meaning of observations, but to facilitate the process of securing them. Thus Hale intended to "apply" science to making more science.

The "applied" knowledge that Hale sought from Baltimore concerned the secrets of making and operating ruling engines. After writing to Ames, and even before getting a reply, Hale wrote directly to Anderson, suggesting that he take leave from Hopkins to oversee the construction of a ruling engine for Mount Wilson.[84] While still in Baltimore, Anderson immediately took the technical problems in hand, and suggested embedding pipes for constant-temperature water in the walls of a sub-basement room envisioned to house a ruling engine. The circulating water would hold the temperature of the room stable to within 1/20 degree, for highly accurate ruling. Wildly optimistic, Anderson asked Hale whether a grating eighteen by twenty-four inches would be too large or too small.[85]

As head of the Hopkins physics department, Ames was kept informed about the negotiations, and soon informed Hale, "Of course the Anderson matter can be arranged."[86] Anderson would have leave from Johns Hopkins, but no permanent move was at first contemplated. As news of Anderson's planned sojourn spread, Ames informed Hale that he wanted to be the first to tell Hopkins president Ira Remsen, who knew the value of instrument production.[87]

An instrument-designer at Mount Wilson, Francis Pease, who worked on major telescope projects undertaken by Hale, soon became involved as well. Hale suggested that Pease help Anderson with technical drawings for a new ruling engine, and Anderson suggested in reply that Pease could visit Baltimore to discuss the existing machines directly. Pease made the trip, and reported back to Hale, "I've examined Rowland's machines pretty carefully and discussed the new machine pretty thoroughly with Anderson so will start in on the full size sketches tomorrow." Within three months after Hale's initial offer to Anderson, Pease had examined the machines and was back at Mount Wilson, making detailed drawings. Anderson's leave from Hopkins was then approved, and Ames generously proposed to split his salary with Hale during the stay in California.[88]

Anderson was in overall charge of initiating the Mount Wilson ruling operation; Pease, a mechanical engineer by training, made the

---

[84]Hale to Anderson, 24 November 1911, GEHM.
[85]Anderson to Hale, 2 December 1911, GEHM.
[86]Ames to Hale, 3 December 1911, GEHM.
[87]Ames to Hale, 2 January 1912, GEHM.
[88]Hale to Anderson, 8 December 1911; Anderson to Hale, 23 December 1911; Hale to Ames, 9 January 1912; Pease to Hale, 21 January 1912; Hale to Ames, 6 February 1912; Ames to Hale, 29 January 1912, GEHM.

plans workable. To perform the function of Schneider, Hale suggested machinist Clement Jacomini: "He takes an artist's pride in his work." As the project advanced and technical problems presented themselves, Anderson reported, "Jacomini's respect and admiration for Rowland and his machine increased by leaps and bounds." Like Pease, Jacomini visited Baltimore, and his "confidence was tempered by an appreciation of the difficulties to be overcome."[89] Secrecy had not been the only impediment to the creation of superior ruling engines outside of Baltimore; sheer technical difficulty was another.

In effect, Anderson would "visit Pasadena at intervals in the years 1912-1915," during which he oversaw the construction of the first Mount Wilson ruling engine. In support, Hale's machine shop added the capacity to cast speculum metal blanks in an on-site foundry, and to ready the blanks for ruling, thus performing the step that Rowland had contracted out to Brashear.[90] By 1915, Hale and Anderson shared the hope that Anderson: who was well aware of the limitations of his home institution, would make a permanent move to Pasadena. In one test of the resolving power of his instruments at Hopkins, he observed two components of a spectral line that were separated by only 0.014 angstrom. He wrote, "I think they are as well separated as might be expected—when it is remembered that our spectroscope is mounted in a 4th story floor, in an old building in the middle of Baltimore."[91]

The question of a permanent move to Mount Wilson was delicate, given Ames's largesse toward Mount Wilson in allowing Anderson to work there during leave. Ames actually hoped that Hale would employ Janet Howell (Johns Hopkins Ph.D., 1913), not Anderson,[92] who was already training a replacement to take over the ruling operation at Hopkins. Hale informed him in 1915 that Jacomini had the Mount Wilson engine ready for trial, and he hoped that Anderson might come permanently: "I often find myself wishing that you were tackling some of our solar or stellar problems. If you ever see your way to come I wish you would let me know." Anderson finally accepted a

[89]Hale to Anderson, 30 December 1911; Anderson to Hale, 5 November 1913; Hale to Anderson, 14 January 1914, GEHM.
[90]Horace W. Babcock, "Diffraction Gratings at the Mount Wilson Observatory," *Vistas in Astronomy*, 29 (1986): 157.
[91]Anderson to Hale, 8 February 1915, GEHM.
[92]Hale declined to employ Howell because of her gender. He wrote, "We have no accommodations for women on the mountain, and it is not practicable for them to live there and carry on observational work." Howell could see the instruments and learn how to use them, but could not carry out regular work there. Hale to Ames, 10 March 1915, GEHM.

Had Howell been married to an astronomer or physicist, however, the situation might have been different. When Hale was trying to attract Charles Mendenhall (Johns Hopkins Ph.D., 1898) to Mount Wilson, and the latter demurred because he was married, Hale told him, "suitable quarters could probably be provided for ladies." Hale to Mendenhall, 5 April 1905, ibid.

permanent position at Mount Wilson in January 1916, and arrived in July. On examining Jacomini's handiwork, he reported, "The ruling machine certainly runs beautifully."[93]

Although ruling expertise was Hale's initial reason for taking an interest in Anderson, the latter predicted that, once in operation, the engine would not require much of his own attention.[94] His role was supervision, not actual operation. Thus, at Mount Wilson, Anderson pursued various research interests. One pertained to the comparison of gaseous vortices in the laboratory with vortices in the solar atmosphere. Before coming to California, Anderson also expressed hope for new formulas interrelating spectral lines and bands. After his arrival, he studied the influence of an electric field on certain spectra (the Stark effect).[95] He also used extremely high voltages and large electric currents to vaporize fine wires, very briefly creating temperatures approaching those in the sun and stars, so that he could record the spectra from the resulting flashes of light, for comparison with solar and stellar spectra.[96]

Anderson's application of physics to Hale's vision of astrophysics was soon interrupted by war. After the entry of the United States into World War I, national interests laid claims on the high precision attained at Mount Wilson. The National Bureau of Standards faced the need to test thousands of gauges each day, and Anderson volunteered the Mount Wilson facilities for the construction of micrometers for the Bureau to use as master gauges, to calibrate other gauges.[97] But this duty diverted shop work away from astronomy. By the end of 1918, after the war, Hale expressed the hope that Anderson and his colleagues would soon return to astrophysical problems. Hale himself had to remain in the East to set the new National Research Council and the International Research Council on a firm footing. As late as 1921, Hale still had to ask the National Bureau of Standards to obtain measuring scales elsewhere, so that he could once again devote the Mount Wilson ruling engines to making gratings, rather than ruling measuring scales.[98]

That Keeler, Hale, and Anderson were all educated in physics suggests that much of the impetus in the new discipline of astro-

---

[93]Anderson gave his replacement's name as Mackenzie. Anderson to Hale, 27 March 1915; Hale to Anderson, 3 June 1915, 19 June 1915, 20 July 1915; Anderson to Hale, 2 August 1915, 6 January 1916, 7 July 1916, GEHM.

[94]Anderson to Hale, 2 August 1915, GEHM.

[95]Hale to Anderson, 19 June 1915; Anderson to Hale, 4 July 1915, 9 May 1917, 22 December 1918, 12 January 1919, 8 April 1919, GEHM.

[96]Anderson, "The Spectrum of Electrically Exploded Wires," *ApJ*, 51 (1920): 37-48.

[97]Anderson to Walter S. Adams, 8 June 1917, in Anderson Correspondence, GEHM.

[98]Hale to Anderson, 30 December 1918; Hale to S.W. Stratton, 5 February 1921, GEHM.

physics came from the terrestrial side, physics, rather than the celestial side, astronomy. Rowland had envisioned the possibility of employing Hopkins instruments in stellar spectroscopy, but in fact this application of gratings did not become routinely successful until 1929 and after, due to modifications in design. This occurred at the hands of investigators other than Keeler, Hale, or Anderson. But the invasion of astronomy by physicists, and the new phenomena revealed by astronomical spectra, support the assertion by Harwit: "New cosmic phenomena frequently are discovered by physicists and engineers or by other researchers originally trained outside astronomy."[99] Given that astrophysics was a new, hybrid discipline, this conclusion is not surprising, but it suggests that disciplinary border-crossings may bear fruit, especially when they bring new ways of seeing.

However beneficial Anderson's move may have been to Mount Wilson Observatory, his departure from Johns Hopkins left the ruling operation there in the doldrums. The ruling engines in Baltimore had been inoperable between 1904 and 1909 due to water damage, but Anderson had put them back in operation, and kept them running until 1916. Then, Hale's coup was Ames's loss, and the Baltimore engines were largely idle until 1923, when Robert Wood took charge of them.[100] After that, physicists on both coasts of the United States held the power to divide and rule, and to keep correlating mundane and celestial light. But, as it turned out, the whole process was haunted, and only hardheaded mechanical engineering at Hopkins could eliminate the outré.

Over the decades and at the respective institutions where they practiced physics, Hopkins physicists continued to interact. They corresponded, consulted together, and shared a common concern in grating production. They were critical users of the instruments; the grating carried the research program with it. Hopkins physicists and colleagues shared a common interest in the mapping and identification of spectral lines, and in the elimination of false lines.

---

[99]Harwit, *Cosmic Discovery*, 20. Original in italics. Harwit lists Fraunhofer as one such outsider.

[100]George R. Harrison, "The Production of Diffraction Gratings. I. Development of the Ruling Art," *Journal of the Optical Society of America*, 39 (1949): 414.

The Ghosts in the Instrument

The hegemony of Rowland's instrument stood largely unchallenged for about twenty years after he invented it. The instrument, Rowland wrote, increased the accuracy of wavelength determination by tenfold or more.[1] When a serious challenge to the veracity of the device arose in 1901, it came from outside, from a physicist at Harvard. Both the challenge and the response, which then remained in conflict for another twenty years, illustrate the dynamics of the physics community in America at a time when new conceptions of the atom arose in Europe. Far from regarding atoms or subatomic particles as their principal objects of study, physicists of the Hopkins school treated the individual spectral line as the most basic of all directly observable phenomena.

From the perspective of the Hopkins school, the greatest impediments to natural knowledge were "ghosts," the name for spectral lines that did not faithfully represent nature, but rather arose from man's imperfect means of separating the discrete, fine shades of color imputed to the elements. Ghosts were reflections of flawed human agency; in its ideal form, that agency left no trace. All the ingenuity that Rowland invested in his ruling engines and instruments was intended to render them unproblematic.[2] The instrument was

---

[1] "On the Relative Wave-Length of the Lines of the Solar Spectrum" (1887), in *PPR*, 512.

[2] As Simon Schaffer observes in regard to Isaac Newton and his prism, makers of scientific instruments strive to establish their devices as "uncontestable transmitters of messages from nature." Schaffer, "Glass Works: Newton's Prisms and the Uses of Experiment," in *The Uses of Experiment*, ed. David Gooding, Trevor Pinch and Simon Schaffer (Cambridge: Cambridge University Press, 1989), 70.

generally obscured by the very light that it successfully dispersed. When it worked properly, it yielded a spectrum accepted as entirely natural; when it was flawed, the resulting spectrum was seen as disturbingly artificial, even unnatural. Hence "ghosts."

The root cause of ghosts lay in the imperfect functioning of ruling engines; a perfect grating would not produce them. If the ruling engine possessed any periodic extraneous motion, this could result in a periodic misplacement of a groove, and the resulting grating would then have not only the intended regular spacing of grooves, but also an additional, unintended periodic spacing effect. The latter would result, in effect, in a very coarse secondary grating superimposed on the first. This secondary grating in turn produced spectral lines that did not belong to the expected spectrum and were categorized as ghosts.

The critique of 1901 came from Theodore Lyman at Harvard, who asserted that Rowland concave gratings produced ghosts in the ultraviolet region of the spectrum. The interaction between grating user (Lyman and later others) and grating producer (the School of Light) displays the conservative, or preservationist, proclivity of the Hopkins school, in its desire to maintain the "rule" of Rowland's ruling engines, and to overcome the faults that Lyman and others detected. (Lyman's observations of the hydrogen spectrum in ultraviolet, after allowance for ghosts, became fundamental to new models of the atom.) While the specter of ghosts haunted attempts to conserve Rowland's engines, modern ideas about the atom continued to cross over from Europe, and intermingle with empirical data from laboratories in the U.S., producing an international mélange.

Although Rowland recognized the importance of a theory of matter, he rarely considered conceptions of the atom, in part because he believed molecules and groupings of atoms to be so complex as largely to defy description. He conveyed this view to his students, as well. He reported telling them: "'A molecule of matter is more complicated a great deal than a piano. Counting the overtones and everything, you would not probably get up anywhere near the number of tones you get out of a single molecule of uranium.'" Rowland said, "when I come to think what a molecule is and try to get up some theory of it, I quite agree . . . that we don't know anything about it."[3] An analogy with acoustics, comparing atoms or molecules with musical instruments, was common to physics of the late nineteenth century, and recurred in the later work of his students. One of the difficulties that Hopkins-school physicists later saw in the Bohr atom of 1913 was

---

[3] "The Röntgen Ray, and Its Relation to Physics" (1896), in *PPR*, 586.

the absence of any vibrating parts.[4] On the analogy with acoustics, any light source would, like a musical instrument, be expected to have vibrating parts.

Except to link specific elements to specific spectral lines, Rowland rarely wrote on matter. But in one of his few references to the subject, he said:

> The theory of matter, then, includes electricity and magnetism, and hence light; it includes gravitation, heat, and chemical action; it forms the great problem of the universe. When we know what matter is, then the theories of light and heat will also be perfect; then and only then, shall we know what is electricity and what is magnetism.[5]

For Rowland, as for many other nineteenth-century spectroscopists, the analysis of light promised knowledge about the constitution of light-emitting matter, but he apparently believed that an ultimate understanding lay in the far distant future. He was still amassing the large corpus of wavelength data that would eventually require explanation.

Rowland's skepticism was consistent over the long term. He had written about the atom in one of his student notebooks for 1868: "It is a *something* of which we cannot conceive and which we have no right to suppose exists. . ... Now it seems to me this is a very unphilosophical way of doing, to go beyond our knowledge in this manner, and set ourselves to work *creating* things." At that date, Rowland hoped for a return to the ideas of Michael Faraday and R.J. Boscovich, on matter as various groupings of forces acting from "one and the same center."[6] Much later, in 1899, he still harbored skepticism:

> The round hard atom of Newton which God alone could break into pieces has become a molecule composed of many atoms, and each of these smaller atoms has become so elastic that after vibrating 100,000 times its amplitude of vibration is scarcely diminished. It has become so complicated that it can vibrate with as many thousand notes. We cover the atom with patches of electricity here and there and make of it a system compared with which the planetary system, nay the universe itself, is simplicity.[7]

---

[4]Frederick A. Saunders (Johns Hopkins Ph.D., 1899) at Harvard complained that "there is no mechanism in our atomic model which appears to vibrate, and some would prefer to leave us no medium in which the vibrations might travel." Saunders, "Some Aspects of Modern Spectroscopy," *Science*, 59 (1924): 50.

[5]"Address as President of the Electrical Conference at Philadelphia, September 8, 1884," in *PPR*, 635.

[6]1868 Notebook, Vol. 2, 42, 43, HAR.

[7]"The Highest Aim of the Physicist," 671.

Physicists coming from the Hopkins school were slow to embrace atomic theory in its various forms. The multitude of spectral lines they observed required ever-greater complexity.

But if Rowland was averse to atomic theory, he was not necessarily averse to all theory, as typical portrayals of American physics in this era might suggest. For example, he was prepared to think abstractly and mathematically about physical phenomena in the case of incompressible fluids and spherical waves of light[8]—and the process of devising and producing gratings was hardly devoid of theory. Both Rowland and his student Joseph Ames explored what they termed the theory of the grating. The former wrote, "This study I at first made with a view of guiding me in the construction of the dividing engine for the manufacture of gratings." He developed the theory years before he published it, and said his original object was "to obtain some guide to the effect of errors in gratings so that in constructing my dividing engine I might prevent their appearance if possible." Periodic errors causing ghosts might, he said, result from human errors creating "'drunken' screws, eccentric heads, imperfect bearings, or other causes."[9]

Rowland had the benefit of others' work on ghosts when he began production in 1882. Just as he relied on the experience of Harvard astronomer William Rogers in building his first ruling engine, he had access to mathematical studies of ghosts carried out by an instructor in logic at Johns Hopkins, the polymath Charles S. Peirce, who taught there from 1879 to 1884. Peirce performed an extensive analysis of ghosts produced by the diffraction gratings of Lewis Rutherfurd, and published it in the *American Journal of Mathematics*.[10] Rowland's colleague in mathematics, James Joseph Sylvester, had started the journal only a year before, with help from Rowland and Simon Newcomb.[11] (Rowland himself also published articles in the

---

[8]Mathematical papers on these two topics do not appear in *PPR*, which a committee published in 1902, because Rowland had opposed republication of his more mathematical papers. The committee included "only the distinctly physical papers." In *PPR*, v. Two of the mathematical articles were: "On the Motion of a Perfect Incompressible Fluid when no Solid Bodies are Present," *American Journal of Mathematics* 3 (1880): 226-268; and "On the Propagation of an Arbitrary Electro-Magnetic Disturbance, on Spherical Waves of Light and the Dynamical Theory of Diffraction," ibid., 6 (1884): 1-23.

[9]"Gratings in Theory and Practice" (1893), in *PPR*, 525 with note, 536. On the "Theory of 'Ghosts'" see ibid., 536-544.

[10]Charles S. Peirce, "On the Ghosts in Rutherfurd's Diffraction-Spectra" (1879), in *Writings of Charles S. Peirce*, ed. Christian J.W. Kloesel, Vol. 4, *1879-1884* (Bloomington: Indiana University Press, 1986), 50-67.

[11]French, *A History of the University Founded by Johns Hopkins*, 51-53. Sylvester told Gilman he considered Rowland to be one of "a corps of no less than eight mathematicians—actual producers and investigators—real working men" at Johns Hopkins. Quoted in Nathan Houser, "Introduction," *Writings of Charles S. Peirce*, xxix.

Hopkins-based *Journal*.)[12] In his analysis, Peirce attempted to relate the ruling of specific gratings to the position of specific ghosts. After comparing his predictions with the measured positions of ghosts, he reported "a happy confirmation of the theory."[13]

The term "ghosts" originally arose in astronomy, where it signified unwanted secondary images seen through a telescope in addition to the primary image being pursued, due to a lens defect.[14] Peirce described ghosts in spectroscopy as "repetitions of the principal spectrum."[15] In fact, diffraction gratings normally gave rise simultaneously to multiple spectra of a source, labeled "first-order," "second-order," and so on, in sequence of increasing dispersion. These spectra arose from interference effects of one wavelength, two wavelengths, and so on. At times higher-order spectra were useful, because of their high dispersion, and spectra of different orders could be compared quantitatively. Ghosts, however, fit none of the expected spectra, and Peirce showed their origin in faulty ruling.

Peirce's interest in wavelengths arose from his researches in metrology, required by his work for the United States Coast and Geodetic Survey. For the Survey, Peirce sought "the precise measurement of a wave-length of light, in order to obtain a check upon the secular molecular changes of metallic bars used as standards of length." This implicated spectroscopy in the pursuit of measures more stable than tangible matter itself, and "proceeded on the assumption that the wave-lengths of light are of a constant value." Peirce hoped that "the standard length may be compared with that of a wave of light identified by a line in the solar spectrum." But it was possible that false lines might skew any measurements of valid lines. He said, "There was doubt as to whether the lines were displaced by 'ghosts,' and this led to the mathematical inquiry." In order to be certain of the wavelength of individual lines, Peirce had to ferret out ruling errors—"the progressive, the periodic, and the accidental." Thus the elimination of ghosts contributed to the maintenance of a standard meter, as later it helped in the definition of a standard atom.[16]

There can be no doubt that Peirce met Rowland during his five years at Johns Hopkins. In 1878, prior to his part-time appointment,

---

[12]"On the Motion of a Perfect Incompressible Fluid;" "On the Propagation of an Arbitrary Electro-Magnetic Disturbance."

[13]Peirce, "On the Ghosts," 64.

[14]*The Oxford English Dictionary* (1978), s.v. "ghosts".

[15]Peirce, "On the Ghosts," 60.

[16]Peirce, "The Width of Mr. Rutherfurd's Rulings" (1881), in *Writings*, 240; "Spectroscopic Studies" (1879), ibid., 4; "Comparison of the Metre with a Wave-Length of Light" (manuscript, 1881-1882), ibid., 269; "Spectroscopic Studies," 5; "Comparison of the Metre with a Wave-Length of Light," 270.

Peirce gave President Gilman a lengthy critique of the new physics department established by Rowland. To address the problems, and probably with his own career in mind, Peirce suggested that the department employ a second full professor of physics, for organization and administration.[17] This appointment did not materialize, but in the lesser position that he actually filled, he had many opportunities to converse with Rowland. "Peirce often saw Rowland at the meetings of the Johns Hopkins Scientific Association and the Mathematical Seminary and he frequented and probably used Rowland's laboratory."[18]

When Rowland later needed an absolute measurement of the wavelength of one spectral line, to anchor his lengthy tables of thousands of wavelengths in the solar spectrum, he turned first to a "determination of Peirce made for the U.S. Coast Survey with Rutherfurd's gratings and not yet completely published." Rowland averaged a number from Peirce, corrected for an error in the ruling of the Rutherfurd instrument; a number from Ångström, corrected for poor definition and an error in taking the length of a standard meter; and a determination by Louis Bell, in his own laboratory in Baltimore.[19] Peirce had advocated the use of one specific line as a standard of reference,[20] and Rowland acknowledged in 1888, "My map of wavelengths is based upon Professor Charles S. Peirce's measurement of a line in the green portion of the spectrum."[21]

As it eventually turned out, the *relative* measurements in Rowland's map were the most enduring, while the *absolute* anchor to which he attached them was most in error, after improved measurements by interferometer. Considering that Rowland expressed wavelengths first to the nearest hundredth of an angstrom unit, and eventually to the nearest thousandth, the absolute or systematic error eventually attributed to his solar tables was enormous: 0.212 Å.[22] However, the relative measurements, with their accidental errors, remained accurate enough that they were "the standard for the first quarter of the twentieth century."[23]

---

[17]For a discussion of Peirce's critique, see Miller, dissertation, 198-205.

[18]Houser, "Introduction," xxix.

[19]Rowland, "On the Relative Wave-Length of the Lines of the Solar Spectrum" (1887), 513-514.

[20]Peirce, "The Width," 240.

[21]Quoted in *Writings of Charles S. Peirce*, 576 note.

[22]The International Union for Cooperation in Solar Research adopted a red line from cadmium as the primary absolute standard of wavelength in 1907. It was this measure that differed from Rowland's system by 0.212 Å. Charles E. St. John, Charlotte E. Moore, Louise M. Ware, Edward F. Adams, and Harold D. Babcock, *Revision of Rowland's Preliminary Table of Solar Spectrum Wave-Lengths* (Washington, DC: Carnegie Institute of Washington, 1928), v.

[23]Hearnshaw, *The Analysis of Starlight*, 155.

Rowland may have been mindful of the possibility of absolute error when he published his first solar tables in 1887 under the heading of "Relative Wave-Length." Indeed, when still a student of engineering at Rensselaer in 1868, he had concluded privately that the absolute was virtually unknowable. He wrote then, "*Everything* in the universe seems to me to be relative." Absolute measures could not apply to motion, time, or space, because measurement always involved relation. Man conceived things only by their relation to each other.[24] (In his mature career, Rowland did not completely eschew absolute measurement: there was an absolute scale in thermometry, and an absolute value of electrical resistance.)

Peirce provided Rowland with an important starting point, in his absolute values of wavelength and his understanding of ghosts. However, like the ruling operations of William Rogers, Peirce's theoretical study of ghosts apparently came to a halt in 1882, when Rowland assumed command of the field. Working mathematically, Rowland soon convinced himself that he understood the possible errors in ruling, and therefore the causation of ghosts. Experience led him to conclude, "The ghosts are very weak in most of my gratings." In the extremely bright solar spectrum, ghosts never appeared. In other spectra, they "never cause any trouble, as they are easily recognized." Rowland knew that ghosts "increase in intensity as compared with the principal line as the square of the order of the spectrum," so that extremely high-order ghosts would be very bright, but he felt that he had the situation under control.[25] This state of affairs lasted for twenty years.

Then came Theodore Lyman. The Harvard physicist was the most exacting user of Rowland gratings, and placed the greatest demands on their performance. Unlike physicists at Johns Hopkins, who were probably personally familiar with the painstaking efforts required to operate the ruling engines, and so reluctant to criticize, Lyman did not hesitate to fault the gratings, once he became convinced that ghosts were, in fact, present. In retrospect, he was an exemplar in the practice of laboratory spectroscopy.

Lyman's name is best known today in association with the most basic spectral series of hydrogen, in the ultraviolet, now seen as arising from electron transitions to the innermost shell of the hydrogen atom. (In the relevant formula, use of the integers one and two yields the wavelength of the first line in the series, or "Lyman alpha.") However, what is most striking about the actual record of Lyman's research, his laboratory notebooks up to 1914, is the relative absence of goals defined by formulas. A directing principle cited more frequently

---

[24]Notebook 2, 1868, Box 20, HAR.
[25]"A Few Notes on the Use of Gratings" (1889), in *PPR*, 519.

was *purity*.[26] The relative absence of formulas was consistent with the tone of physics research at Harvard in that day, even though Lyman took up research in the extreme ultraviolet in 1897, after the Balmer and Rydberg formulas, but long before the Bohr atom.[27]

Lyman labeled the false lines he discovered as "phantoms," in contrast to the "ghosts" already detected by Rowland and Peirce. "Ghosts" arose from small-scale variations in the ruling of gratings; phantoms arose from variations that were not small-scale and therefore not readily susceptible to mathematical analysis.[28] (In the long run both types of lines were called ghosts.) Some twenty years after Peirce's complex calculations, Lyman cut the gordian knot regarding phantoms by showing that certain groups of spectrum lines, resulting from a light source refracted through a prism before striking a Rowland grating, were displaced in an unusual way. Prisms displace light of different wavelengths by different amounts, but in one experiment, a prism gave equal displacement to separate lines supposedly representing different wavelengths. Therefore phantoms were *extra* lines representing a single wavelength, and were not truly distinct.[29]

Working with the magnesium spectrum, Lyman studied three groups of lines that were "strikingly characteristic and identical in appearance." Were all of the three groups, he asked in 1901, "real first spectrum lines"? Lyman interposed the prism between the light source and the grating, "to produce virtual images of the slit whose displacements were nearly proportional to the refrangibility of the light." But the three groups were then displaced by *equal* distances. Lyman therefore concluded that all three groups were caused by incident light of one and the same wavelength. Mathematical analysis could not have unmasked the pretenders: "The wavelengths of the true group and its phantom repetitions bear no simple relation to one another."[30]

Lyman thus vindicated his nonmathematical approach to physics and also demonstrated that Johns Hopkins University would not have the last word in spectrum analysis.[31] The kind of perfection sought by

---

[26]Theodore Lyman, Laboratory Notebook 1900-1906, 19 April 1904; Laboratory Notebook 1910-1920, 17 January 1914, Theodore Lyman Papers (TL), Harvard University Archives, Cambridge, MA.

[27]Katherine Russell Sopka, in her history of quantum theory in the U.S., was surprised to find no citations of Lyman's work in support of Niels Bohr's theory of the atom immediately after it emerged in 1913. Sopka, *Quantum Physics in America: The Years Through 1935* (New York: American Institute of Physics, 1988), 53 note.

[28]Lyman, "False Spectra from the Rowland Concave Grating," *The Physical Review*, 12 (1901): 8.

[29]Ibid., 3-4.

[30]Ibid., 3-5.

[31]At the same time, Lyman's approach conformed to Rowland's dictum that nature and truth are to be found in the *laboratory*, more than in the theoretical domain.

Rowland, he implied, could not possibly have been achieved. Rowland's gratings had been criticized before; Rowland himself had analyzed some causes of ghosts, and was not unresponsive to critical appraisals. But by and large he had been able to establish, by machining and by analysis, the trueness of his gratings. Lyman demurred that absolute equality of width and spacing in grating lines simply could not exist, "in view of the very minute distances involved and the almost inconceivable rigidity of the ruling engine which would be necessary."[32]

Lyman dismissed mathematical arguments in advance by saying that the departure from equal spacing did not invite analytic discussion, because it was "an experimental fact" that the phantom spectral lines were "not harmoniously placed." He also held that false lines were "most clearly visible" in the region he studied, the extreme ultraviolet, "owing to the absence of strong real lines." Thus anyone who would disagree with Lyman faced a very high cost of dissenting,[33] in the construction of a spectroscope for the ultraviolet. The ringing criticism of Rowland's instrument, first among Lyman's conclusions, resonated with Lyman's laboratory aims: the first-order spectrum obtained was simply "not pure."[34] There was hope, however, at higher magnification. False lines were fainter from gratings of larger radius. From a 21-foot grating (Lyman's department possessed one), false lines would, for the most part, not be significant.[35] Thus Lyman's conclusions strengthened the logic that demanded high dispersion of the spectrum.

Eventually, Lyman sought an explanation of false lines, in mathematical terms, in the work of the German physicist Carl Runge. Lyman had tried to use Rowland's formulas to discover the cause of the new false spectra, without success. Lyman concluded that, while previous analyses allowed for a periodically varying displacement of *all* grooves in a grating, the false lines he observed could be explained by the displacement of *one* out of every *n* grooves.[36] The mechanical cause, the bug, which led to this error, remained a mystery for years.

Only with falsity isolated to his satisfaction could Lyman pursue truth. A program of "normal" science began for Lyman only after he had winnowed out the false spectral lines and cleared the way, in his

---

[32]Lyman, "False Spectra," 9.

[33]Bruno Latour discusses some of the ways in which discussion or dissent may require the accumulation of costly resources. Latour, *Science in Action* (Cambridge: Harvard University Press, 1987), 69-70.

[34]Lyman, "False Spectra," 9, 13.

[35]Ibid., 12.

[36]Lyman, "An Explanation of the False Spectra from Diffraction Gratings," *The Physical Review* 5 (1903): 257-258, 265-266. Lyman found Rowland's treatments inadequate even though Rowland had in fact considered the possibility of "one groove in *m* misplaced." Rowland, "Gratings in Theory and Practice," 535.

own estimation, to making valid observations of nature. His routine use of the 21-foot-radius grating was part of the solution. His principal years of research lasted until the First World War. Lyman later recalled: "My most productive years lie between the autumn of 1902 and the spring of 1917."[37] (In 1901-1902, he studied at the Cavendish Laboratory in Cambridge, England and in Göttingen.) Throughout this time, his primary research specialty remained the same, spectroscopy of the extreme ultraviolet, and he believed one of his own principal attributes was doggedness: "The work is characterized rather by diligence than by any great originality of conception. It was a new field and I followed my nose without interruption from outsiders."[38] Lyman's publications on ultraviolet spectra were typically much shorter and less analytical than his tracts on Rowland's instrument. They were simply reports of a few numbers.

With his spectroscope validated, or at least with its shortcomings noted, Lyman used the same basic laboratory setup, with occasional small adjustments, for years. His light source was generally a gas made luminous by electrical discharge. His spectroscope was evacuated to a very low pressure (a near-vacuum), in order to avoid the absorption caused by air. The purpose of each run was generally to take a photographic plate of lines in a given section of the ultraviolet spectrum. Lyman's work might be seen as the stockpiling of photographic plates—from which wavelength values could be derived.[39] Lyman was not extremely concerned with the generation of formulas that would compactly specify groups or series of lines. He was more concerned with attributing specific spectral lines to specific elements than with new levels of precision measurement.

After ghosts, and in contrast to Rowland's drunken screws and eccentric heads in the machine shop, Lyman's principal opponents in the laboratory were leaks, impurities, and "fog." The vacuum in the spectroscope was never perfect, and Lyman often recorded not only the (low) pressure in the apparatus, but also the rate of leakage. Air was an impurity, and its spectrum had to be distinguished from that of hydrogen and other gases being observed. Thus, in order to isolate the desired spectra, Lyman needed to know the spectrum of air. Although most of his observations could only be made from photographic plates, ultraviolet light being invisible to the eye, he saw that the spectrum of air was "white-ish" compared to the "very pink"

---

[37]Lyman, "Biographical Notes" (1938), 3, TL.

[38]Ibid.

[39]Latour has described laboratory activity in science generally as the amassing of "inscriptions," of which wavelengths might here be an instance. *Laboratory Life: The Construction of Scientific Facts* (Princeton: Princeton University Press, 1986), esp. 47-49.

characteristic of hydrogen.[40] Lyman also attempted to record the air spectrum, but on at least one attempt he failed: "No result except Fog." In addition to its contaminating lines, air also tended to block ultraviolet light. In the attempt, pressure in the apparatus was "too high for transparency."[41] Lyman had already found that air in the spectroscope, even at very low pressure, "absorbs everything below 1600."[42] Thus there were two reasons for obtaining a good vacuum: purity and transparency. In practice, isolation of hydrogen and other elements, and the exclusion of contaminants, demanded much attention and innovation.

For unwanted lines in the spectrum, another possible source was the electrodes in the luminous discharge tube. Lyman shifted between silver, copper, and aluminum electrodes to rule out any metallic contribution to the spectra.[43] Trace gases in the discharge tube were also a concern. To eliminate them as far as possible, Lyman flushed the apparatus, as in April 1904: "Washed pumped & rinsed with H about six times."[44] In the same way that trueness and precision, in machining and measurement, were guiding ideals to Rowland, purity, which ruled out falsity, was a consistent aim of Lyman. Rowland wanted lines in great abundance; Lyman wanted them relatively few, as they were in the ultraviolet region, when pure gases were observed through a vacuum. (This could still entail hundreds of lines, in total, over time.)

In 1914, as Lyman's work gained importance outside Jefferson Laboratory, he devised a new means of providing hydrogen to the discharge tube. His steps resembled chemical engineering as much as modern physics. The supply tube ran through liquid air, acting as a trap for any air impurity, which would condense there. It also ran through four vertical chambers containing sodium hydrate. Lyman believed he had increased the purity of the hydrogen.[45]

In each run, the goal was generally a good photographic plate, but there were many possible impediments. A principal obstacle was fog, a general darkening of part or all of the plate, resulting from diffuse ultraviolet light in the apparatus, faulty preparation of the plate, or other reasons. Thus diffuse light was another impurity to be eliminated in obtaining distinct spectra, and this involved the makeup of the spectroscope. In 1913 Lyman tried a grating manufactured by

---

[40]Lyman, Notebook 1900-1906, 27 April 1904, TL.

[41]Ibid., 28 April 1905.

[42]Ibid., 19 April 1904. In a subsequent entry, he found that air was transparent again for wavelengths less than 1160 Å. Ibid., 24 June 1904.

[43]Lyman, Laboratory Notebook 1910-1920, 30 December 1910, 19 May 1911, TL.

[44]Lyman, Notebook 1900-1906, 27 April 1904, TL.

[45]Lyman, Notebook 1910-1920, 17 January 1914, TL.

Michelson at Chicago, but diffuse light was a problem: "Test with Al spark seemed to show that Michelson grating yielded nearly as strong a spectrum as my grating. Causes of failure, pronounced scattering in the Michelson grating producing an obscure back ground, lack of sharp adjustment, opacity of qtz."[46]

Lyman only occasionally changed the configuration of the spectroscope, giving more attention to the preparation of samples. Hence it is noteworthy that in December of 1913 he recorded, "Spectroscope opened + *grating turned* to bring region l 1400 on red end of plate."[47] In other words, he was now pursuing spectral lines with wavelengths shorter than 1400 Å. This opened a new, aggressive period of extending research further into the ultraviolet, during which Lyman reported detection of lines at 1216 Å and 1026 Å—the very spare results which, seen in retrospect, have given him such a significant place in the history of atomic physics.

In a short letter to *Nature,* Lyman reported that he had improved on the prior work of Schumann, based on "an improvement in technique consequent on an experience of ten years."[48] His patience in acquiring definitive measurements was extreme. He once wrote: "But let us remember the fact, established by the history of science, that time and effort spent in the accurate and intelligent observation of nature is never spent in vain."[49] The problem of impurities still bedeviled the experiments, and Lyman still foresaw a need to account for them, but in the detection of lines at 1216 Å and 1026 Å, he wrote, "it may be stated with some degree of certainty that the diffuse series predicted in this region by Ritz has been discovered." (Walter Ritz, a German, developed mathematical expressions for spectral series, including generalizations of the Balmer formula.) There is no mention of atoms or electrons or ground states, nor of corresponding integers. Lyman simply notes that the new series, represented thus far by the two lines, "bears a simple relation to Balmer's formula."[50]

Lyman's approach was, characteristically, phenomenological, omitting concepts of atomic cause and emphasizing observables instead. To him the principal laboratory achievement, clearly, was the extension of observations further beyond one recorded end of the spectrum; detection of lines fitting a modified spectral series formula

---

[46]Ibid., October 1913. "My" grating was a Rowland instrument.

[47]Ibid., 4 December 1913.

[48]Lyman, "An Extension of the Spectrum in the Extreme Ultra-Violet," *Nature,* 93 (1914): 241.

[49]Lyman, "The Spectroscopy of the Extreme Ultra-Violet," *Journal of the Franklin Institute,* 201 (1926): 562.

[50]Lyman, "An Extension of the Spectrum," 241.

was secondary to that.[51] In the previous year, Niels Bohr had restated Ritz's generalized formula in his foundational essay, "On the Constitution of Atoms and Molecules." Bohr demonstrated how the formula might arise from his atomic model, and suggested the occurrence in hydrogen of an ultraviolet series and a second infrared series, "which are not observed, but the existence of which may be expected."[52] Although some connection between Lyman's research and Bohr's theory of atoms is often assumed, in 1914 Lyman did not cite Bohr. And soon afterward, Robert Millikan stated, "Lyman's discovery, subsequent to the birth of the Bohr atom, of an ultra-violet series of hydrogen lines . . . is not to be regarded as a success of the Bohr atom, but merely as a proof of the power of series relationships to predict the location of new spectral lines."[53] (In his 1913 essay, Bohr was only able to cite two thus-far observed series of hydrogen: the Balmer series, and the first infrared series, which Friedrich Paschen had observed by then.)[54]

As Lyman continued to extend his spectrum photography further into the ultraviolet region, he often found helium a useful source of light, because of the ultraviolet emissions it furnished. He believed he had obtained "considerable purity," and had made his vacuum spectroscope "more nearly air-tight than ever before." He thus extended his photography from 900 Å to 600 Å. In a published report he also confirmed hydrogen lines as predicted by Ritz, not only at 1216 Å and 1026 Å, but also at 972 Å. However, there were still puzzling lines recorded from mixtures of hydrogen and helium, which Lyman could not attribute decisively to one element. Despite advances, Lyman laid the blame for the lack of verdict on "the great difficulty of obtaining the gas content of my spectroscope absolutely free from impurities."[55] That certain lines appeared distinctly in a *mixture* of the two gases also delayed a definitive attribution. Still, Lyman found the extension to 600 Å "[w]onderful," and his Harvard colleague Edwin Hall, a Hopkins Ph.D., wrote, "I am delighted that you have got a little nearer the end of the rainbow and I hope you will find the pot of gold that is supposed to be there."[56]

---

[51]In Lyman's personal travels there was an intriguing parallel to his photographic capture of previously unknown spectral lines in the little-known ultraviolet. He took extensive hunting trips, sometimes to remote areas such as Korea, Alaska, British East Africa, Siberia, and Mongolia. Lyman hunted alone in Mongolia, for large game, and brought back a gazelle of a previously unknown species. P.W. Bridgman, "Theodore Lyman," *BMNAS*, 30 (1957): 245-246.

[52]Niels Bohr, "On the Constitution of Atoms and Molecules," *Philosophical Magazine*, 26 (1913): 9-11.

[53]R.A. Millikan, "Radiation and Atomic Structure," *The Physical Review*,10 (1917): 201.

[54]Bohr, "On the Constitution," 9.

[55]Lyman, "A Further Extension of the Spectrum," *Nature*, 95 (1915): 343.

[56]Lyman, Notebook 1910-1920, 20 March 1915, TL. Edwin Hall to Lyman, 7 April 1915, Director's Correspondence: Theodore Lyman, 1910-1938 (DCL), Harvard University Archives, Cambridge, MA.

By 1917, Lyman was able to report, in a characteristically short communication, that he had obtained spectrum photographs as far as 510 Å.[57] However, the year 1917 marked a turning point in his research career. By Lyman's own account, the principal agent of change was the First World War, but something else also touched his work: by the late 1910s, formulas and theory played such a significant role in spectroscopy that they began to guide his research. This was consistent with the vicissitudes of the discipline. A historical survey of spectroscopic articles in the *Physical Review* over its first century found that new ideas "cast the field into a more theoretical vein" by the 1920s, even though theory required "testing, evaluating, readjusting, reconciling and, in some cases, abandoning."[58]

Lyman entered military service in 1917, and before he was "mustered out" in 1919, he had been in command of more than a thousand men. During the war he pursued flash ranging of artillery. Back in Cambridge afterward, Lyman wrote, "I found it impossible to resume my academic work or my research with any enthusiasm. It took some time for me to get back into harness. Meanwhile a number of people had entered the field of vacuum spectroscopy. I had lost the leadership and I never regained it."[59] (Robert Wood and Robert Millikan became important leaders in ultraviolet research.) Lyman was still well respected. Given his position, it could hardly be otherwise: he served as director of Jefferson Laboratory from 1910 until 1947.[60] He continued to publish on ultraviolet spectroscopy, especially of helium. His published output in the 1920s was not significantly less than in the 1910s.

After the war, Lyman gave greater prominence to formulas, especially one proposed for helium:

$$\sqrt{} = 109750 \left(\frac{1}{(\frac{n_1}{2})^2} - \frac{2}{(\frac{n_2}{2})^2} \right)$$

Lyman investigated the case of $n_1=2$ ($n_1=3$ and $n_1=4$ were known), but before the war reported no success. In late 1919, however, he used "a powerful disruptive discharge" to obtain two ultraviolet lines, approximating the formulaic results, at 1640.2 and 1215.1 A. Possible detection of a third line was obscured, "probably due to an impurity."[61] The important distinction from Lyman's earlier work was that here success was no longer defined solely in terms of pushing observations

---

[57]Lyman, "The Limit of the Spectrum in the Ultra-Violet," *Science*, 45 (1917): 187.
[58]Charles Bazerman, "Modern Evolution of the Experimental Report in Physics: Spectroscopic Articles in *Physical Review*, 1893-1980," *Social Studies of Science*, 14 (1984): 187.
[59]Lyman, "Biographical Notes," 4, TL.
[60]Bridgman, "Theodore Lyman," 239.
[61]Lyman, "A Helium Series in the Extreme Ultra-Violet," *Science*, 50 (1919): 481.

further into the ultraviolet, but also in terms of satisfying the predictions of a formula. Lyman's research had become formula-directed. (He continued to extend the ultraviolet. In 1924, working with helium, he reached 256 Å.)[62]

Hopkins professor Wood emphasized another element, sodium, greatly extending the recorded Balmer series for that element. Sodium was salient because of the prominence of the sodium "D" lines in the solar spectrum, and the importance consequently attached to sodium in any understanding of solar radiation. The Balmer series was, in Wood's work, exhibited very completely. According to Wood, only seven lines in the sodium Balmer series were known prior to his own work. Through careful control of experimental conditions, however, "the number of lines in the series was raised to 48, the largest number ever found for any substance exceeding even that shown by hydrogen in the sun and certain stars, which shows about 30 lines."[63] This achievement illustrates a common characteristic of Wood's work, and also one characteristic of spectroscopy of the period. As to the former, Wood's flair for the dramatic is evident. Incremental progress was not sufficient; the results had to amount to a leap forward. As to the latter, cross-comparison of laboratory and celestial spectra was a common constitutive element of spectroscopy, even for non-astronomers. The celestial offered a benchmark, which Wood exceeded, in terms of the number of lines observed.

Except for the first lines (the D lines), the series lay in Lyman's region, the ultraviolet. As well as the proper photographic medium, Wood's study also required high spectral resolution as obtained through high dispersion. The space between lines diminished greatly near the "head" or series limit, becoming extremely narrow for the last 22 lines.[64] A distinct photograph of the series could only be achieved with superior instrumentation, of the kind controlled by the Hopkins school. Wood observed the extended spectrum in absorption rather than emission. He found that, for absorption, greater vapor densities enhanced the Balmer series, but the condition was demanding: "Much time was lost at first, as a result of the chemical ferocity of sodium vapor at high temperatures."[65] This research, like Lyman's, owed something to chemical engineering. (Wood had pursued graduate study in chemistry at Baltimore, Chicago, and Berlin before finally switching to physics.) Wood stated that a further extension of the

---

[62]Lyman, Notebook 1920-1926, 30 October 1924, TL.
[63]Wood, *Physical Optics* (New York: The Macmillan Co., 1924; reprinted from 1911), 443.
[64]Ibid., 444.
[65]Wood, "The Optical Properties of Metallic Vapors," in *Lectures Delivered at the Celebration of the Twentieth Anniversary of the Foundation of Clark University* (Worcester, MA: Clark University, 1912), 98.

Balmer series in sodium was "only a question of a more powerful spectroscope and greater vapor density." In Zurich in 1914, Wood increased the number of measured lines in the series from forty-eight to fifty-eight, but publication of the results was delayed by war.[66]

Only after the First World War did Wood turn to the most fundamental atom, hydrogen. During the war, Wood had experimented with ultraviolet light as a means of secure military communication. Afterward, he again investigated ultraviolet light, because the hydrogen Balmer series led there. Once again, Wood was concerned with experimental procedures and not with atomic models. Once again, too, Wood remained aware of celestial sources—now, nebulae—as benchmarks against which to appraise his laboratory spectra. And once again, the extended spectrum was obtained in part through ingenuity of a chemical kind. Only a sample properly prepared would exhibit extended terms of Balmer's 1885 formula.[67]

To hold the hydrogen sample, Wood used a narrow glass vacuum tube, two meters long in the critical portion. He sighted the spectrograph through bulbs at either end, which were created for clear end-on viewing, and along the axis of the tube, to include as much gas as possible. Wood's apparatus sometimes operated for several days, with hydrogen flowing through, before sufficient purity of the hydrogen sample existed. The flow of gas carried off hydrocarbon vapors and cleansed the inner walls of the tube. When the apparatus arrived at sufficient purity, Wood found that an increase of electrical current through the gas would greatly brighten the Balmer series. The brightness of successive lines diminished markedly with series number, however. The fourteenth line was only 6.7×10^{-5} as bright as the first, and the twentieth line, the furthest captured, was only 1.2×10^{-6} as bright. Successful photography of these lines required minimization of all extraneous lines not belonging to the Balmer series, which would obscure faint Balmer lines. These laboratory studies were more delicate than celestial studies, because from celestial nebulae the fourteenth line was fully 0.02 times as bright as the first, 17,000 times better than the laboratory source.[68]

Unlike Lyman, Wood found that absolute purity of his sample was not greatly to be desired. The Balmer series appeared most

[66]Wood, "The Optical Properties of Metallic Vapors," 98, 111-112. Wood and R. Fortrat, "Principal Series of Sodium," *ApJ*, 43 (1916): 73, 77-78, 80. Wood and his collaborator did not use a grating to achieve this further extension. Rather, the observations were "made in the laboratory of Professor P. Weiss, at Zurich, with the largest and most powerful quartz spectrograph in the world." Ibid., 73.

[67]Wood, "Hydrogen Spectra from Long Vacuum Tubes," *Philosophical Magazine*, 42 (1921): 729-745.

[68]Ibid., 729-730, 732-733, 735.

strongly when the tube contained a trace of water vapor, which led Wood to certain conclusions. Before hydrogen would exhibit the Balmer series, he reasoned, it had to break down from diatomic to monatomic, because molecular hydrogen could not give the desired series. Electric current tended to break diatomic molecules apart, but the effect only occurred in the presence of a trace of water vapor. Without that vapor, the Balmer lines disappeared; in the presence of the vapor, a stronger electric current rendered more of the gas monatomic, and the Balmer series became more distinct. (There was a limit to the brightening, however, which Wood attributed to some unavoidable minimum of diatomic hydrogen.)[69]

The action of water in bringing out the Balmer series became "[o]ne of the most discussed problems in spectroscopy." Soon, but not immediately, Wood provided an explanation. He reasoned that the glass walls of the tube catalyzed the recombination of monatomic hydrogen into the molecular variety. The presence of water vapor on the tube walls prevented contact between hydrogen and the glass, and so prevented the catalysis as well, keeping the sample monatomic. Wood concluded: "The mystery of why the long tube gives a pure Balmer spectrum at the center now appears to be explained."[70] Wood produced a corresponding explanation for the presence of the extended Balmer series in celestial sources such as the solar corona: "The luminous gases are in these cases not in proximity to catalyzing surfaces, and consequently atomic hydrogen of 100 per cent. concentration can exist."[71]

As spectroscopists extended the performance limits of their apparatus, much of their intellectual energy still centered on rendering their spectra more natural, by eliminating the unnatural ghosts. One important figure in this pursuit was a younger spectroscopist, William Meggers. Already a physics instructor in 1914, he read Bohr's paper and decided to devote his career to spectroscopy.[72] He received a Ph.D. from Johns Hopkins in 1917, and spent his career at the National Bureau of Standards. Where the more senior Wood and Lyman tended to restrict their observations to specific elements and spectral regions, Meggers kept the spectra of a great many elements within his purview, in a career that lasted until the 1960s. Unlike Lyman, Meggers was not a teacher, and he was also far less interested in showy demonstrations than Wood.

---

[69]Ibid., 745.0

[70]Wood, "Atomic Hydrogen and the Balmer Series Spectrum," *Philosophical Magazine*, 44 (1922): 539, 541-542.

[71]Ibid., 543.

[72]Paul D. Foote, "William Frederick Meggers," *BMNAS*, 41 (1970): 319.

Meggers soon became caught up not only in laboratory spectroscopy, but also in its implications for the articulation of atomic models. Already in 1920, he thought his laboratory work would secure for him the Nobel Prize.[73] Meggers and a collaborator, Paul Foote, studied by prism the spectral emissions of cesium upon excitation by high-velocity electrons. They found that, for electrons accelerated by a field anywhere between 1.5 and 3.9 volts, cesium atoms emitted "a single-line spectrum rather than a single-series or group spectrum." This "single-line spectrum" was actually the cesium doublet, 8521 Å and 8943 Å, at the start of a principal series. The doublet appeared with maximum intensity at 3.9 volts, which Meggers and Foote held to be the ionization potential for the light-emitting electrons. At higher voltages, other electrons became involved, and a more complex ("group") spectrum appeared.[74]

To situate their research, Meggers and Foote presented "a schematic representation of the caesium atom," which depicted atomic states as vertical lines on a horizontal scale, positioned according to the logarithm of wave numbers. (See Figure 6.) The authors derived the mode of depiction from Raymond Birge, and felt that, "regardless of the theory of atomic structure, it affords a precise picture of the possible series lines in the spectrum of an element—a much clearer picture than may be gained by one unfamiliar with series notation from an examination of the series formulae." From the diagram, energies of spectral lines, and concomitantly their wavelengths, could be derived. The authors elaborated: "On the basis of the Bohr theory, the lines shown in the diagram are really portions of the elliptical or ring orbits about the nucleus." The lowest energy level portrayed was labelled "1.5s," considered the "outermost stable orbit" and also the "innermost unstable orbit," with the same diameter as a normal cesium atom. Bohr's was the only significant theory applied, and interpretation of the lines derived from it. Meggers and Foote concluded that the observed appearance of the doublet, and then the more complex spectrum, "is readily explainable on the basis of Bohr's theory—in fact it affords a strong argument for this theory."[75]

The work never brought Meggers a Nobel Prize, but he became a very active member of a network of physicists, German as well as American, who constantly exchanged materials and insights, and he

---

[73]"Foote and I have cooperated on the production of a scientific paper which will surely win us the nobel [sic] prize and election to our National Academy. It is on the single line spectrum." Meggers to Keivin Burns, 22 January 1920, William F. Meggers Papers (WFM), Niels Bohr Library, American Institute of Physics, College Park, MD.

[74]Paul D. Foote and W.F. Meggers, "Atomic Theory and Low Voltage Arcs in Caesium Vapour," *Philosophical Magazine*, 40 (1920): 96.

[75]Foote and Meggers, "Atomic Theory," 80-81, 96.

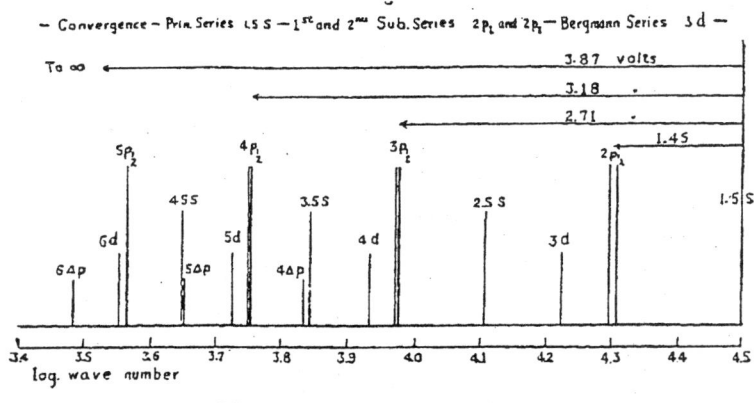

Schematic representation of cæsium atom.

FIGURE 6. *Representation of the energy levels of a cesium atom. From Foote and Meggers (1920)*

played a prominent role in stabilizing internationally agreed-upon wavelength values. From the 1920s, he firmed up the discipline. Also, within the U.S. community, he soon provided motive force, in addition to Lyman's, for the elimination of certain nineteenth-century ghosts still haunting the field, for spectroscopic research remained inextricably entwined with the validation of instruments. Meggers and Lyman held a common interest in the need for unproblematic tools.

Meggers soon launched a vast program of cataloging elemental spectra, including those of relatively obscure elements, but Lyman's agenda for extending the known ultraviolet held more allure for certain other physicists. Two who took up Lyman's mode of research directly were Robert Millikan at the University of Chicago, and Robert Wood at Johns Hopkins. Lyman welcomed both Millikan and Wood to ultraviolet research. Addressing Millikan in 1919, Lyman sounded the same warning he had given for twenty years:

> There is one source of error, however, that should be borne in mind namely "ghosts." If you have been obliged to make rather long exposures, say an hour or more, the danger of "ghosts" is a very real one and the result can never be regarded as certain unless it is confirmed by observations with another grating.[76]

Millikan heeded Lyman, and soon reported extension of the recorded spectrum to 270 Å. Millikan also soon reported to Lyman that ruling engines at the University of Chicago could make gratings to order.[77]

---

[76]Lyman to Robert Millikan, 15 September 1919, DCL.
[77]Millikan to Lyman, 2 October 1919, 30 November 1920, DCL.

In 1921, Lyman complained to a colleague at Johns Hopkins that the Chicago ruling engine was the only engine in operation in the world, and that he found the situation unsatisfactory.[78] It soon changed. Millikan moved permanently to Caltech in Pasadena in 1921, and became chairman of the executive council there as well as director of the physics laboratory.[79] (Millikan, with Ira Bowen, then a graduate student he brought from Chicago, undertook a major program of ultraviolet research in the 1920s.) Pasadena became a center of grating production, not directly at the hands of Millikan, but at those of John Anderson, who had made a permanent move to Mount Wilson Observatory in 1916. Anderson's name would have been familiar to Lyman. (The production of gratings in Pasadena was one of many inducements for Millikan to move to Caltech.)[80] Anderson's departure from Hopkins had left a slack in the production of gratings there that was not taken up until the 1923, when Wood finally took control. There were then two Rowland-derived manufactories, at Baltimore and at Pasadena, as well as the Michelson-initiated operation, with limited production, in Chicago.

In the early 1920s, ghosts still haunted high-dispersion, high-precision spectra, and untangling the spectral secrets of the atom necessarily required the elimination of spectral lines that might be artifacts of a flawed instrument. Rowland ghosts, involving short-period errors in ruling, were fairly well understood, but the origin of Lyman ghosts, seemingly more sporadic, was vexingly elusive. Lyman himself had largely rested his case after finding experimental ways to differentiate between "real" and "false" lines, and making a limited application of theory developed by the German physicist Carl Runge. However, those physicists belonging more properly to the Hopkins school inherited the problem that Lyman had posed too late for Rowland to solve. (Both Rowland and his machinist, Theodore Schneider, died in 1901, the year of Lyman's original critique.) Anderson, at Pasadena, continued to dismantle his engines conceptually in an attempt to isolate sources of periodic error. Hopkins-trained William Meggers, pursuing high-precision research at the National Bureau of Standards, continued to sound the tocsin about ghosts in general, as Lyman had. Runge remained the respected theorist of grating errors. But it would ultimately be Wood, who after 20 years took over the Hopkins operation left by Rowland, who would identify the baffling mechanical cause.

---

[78]Lyman to Joseph S. Ames, 29 April 1921, DCL.

[79]"Strictly speaking, Millikan never served as president." Judith R. Goodstein, *Millikan's School: A History of the California Institute of Technology* (New York: W.W. Norton, 1991), 94.

[80]Babcock, "Diffraction Gratings at the Mount Wilson Observatory," 157.

Holding most gratings in high esteem, Meggers wrote that "it seems almost disrespectful to mention the errors into which they sometimes lead spectroscopists." But he called for "extreme care and watchfulness to avoid spurious results." Almost with the tone of a moralist, Meggers and an associate felt they had to "warn all users" against the "insidious" lines that were a product not of nature but of the limits of human skill in the engineering of optical instruments. Meggers told a colleague, "The discovery of spurious lines among some of our tables have [sic] caused me considerable grief."[81] Meggers requested the pedigree of at least one of his gratings from Johns Hopkins. Each grating had an identifying number, and records were kept on the production of each. In 1919, Meggers's interest was grating #222, tested by Anderson in 1913. Meggers wrote to Ames: "If you can find information in the Grating Ruling Record Book about temperature variations or mechanical causes of periodic errors. . . this may help explain the ghosts."[82]

Meggers's style was experimental, but he reasoned that the only way to identify ghosts positively was by deriving "their more or less complex numerical relationships to the true or parent lines." Rowland ghosts possessed only slight misplacement; Lyman ghosts were much farther from the mark. Meggers was able to find mathematical relationships between Lyman ghosts and the "parent" or unproblematic lines, but not causes: "No theory thus far advanced explains in a satisfactory manner the possible existence of false lines occurring at considerable distances from the parent." Without such a theory, falsity could not be eliminated.[83]

As happened so often in the rise of the related field of quantum physics, Americans turned to a German physicist for theory—in this instance Runge. Meggers conveyed his experimental findings to Runge, who "promptly furnished a complete analysis which explained the Lyman ghosts as due to a double periodic error."[84] Runge was able to derive the position of the particular Lyman ghosts observed by Meggers after assuming, for grating #222, one periodic error at every 298 grooves and another at every five grooves. Lyman ghosts from other gratings arose from errors of different periods. Runge suggested that little had changed in twenty years: "This is the same explanation that I gave to Mr. Lyman in 1902." The physical causes, in some "periodic arrangement" in the ruling engine,

---

[81]W.F. Meggers and C.C. Kiess, "False Spectra from Diffraction Gratings: Part I. Secondary Spectra," *Journal of the Optical Society of America and Review of Scientific Instruments* 6 (1922): 417-429. Meggers to Ames, 3 May 1919, WFM.

[82]Meggers to Ames, 1 April 1919, WFM.

[83]Meggers and Kiess, "False Spectra," 418, 427.

[84]Ibid., 427-428.

remained in question, but "an examination of the machine should be able to tell."[85]

Meggers, trying to manage a resolution of the dilemma in 1922, then forwarded the problem to Anderson at Pasadena, who had the greatest knowledge of ruling engines. Anderson itemized all the components of the engine that could affect the spacing between grooves. He found that the ruling of ghost-free gratings was quite within the realm of possibility, if parts were completely rigid, but deformation of parts, comparable in magnitude to the wavelength of light, was likely. Before actual ruling, the engine had to be warmed up: "the force required to move the carriage increases rapidly at first becoming constant only after several hours of operation." The drive screw was still a critical component, and its actual dimensions depended on the force applied to it. At a typical force, "the compression of the screw when the nut is at its middle is about one half a wave-length of light." But deformation, if constant, was not a problem, and it would remain constant as long as the motive force, F, remained constant.[86]

Anderson looked elsewhere in the machine for possible periodic variation in F, and found a candidate in the drive belt:

> "The belt exerts an upward pull on the machine which will vary somewhat if the belt is not uniform in thickness, owing to the fact that the driving pulley is small, being only two inches in diameter. This varying pull with its resultant distortion of the frame of the machine may perhaps be sufficient to cause the required change in F."

(Meggers had suggested to Anderson that the drive belt was a possible source.) Anderson believed that this effect explained the five-line periodicity. To explain the second periodic error, Anderson proposed small "variations in the length of the belt due to stretching with use." Anderson concluded by noting that Rowland had warned of effects from the drive belt.[87] (Rowland wrote in 1893: "So sensitive is a dividing engine to periodic disturbances that all the belts driving the machine must never revolve in periods containing an aliquot number of lines of the grating; otherwise they are sure to make spectra due to their period.")[88]

In 1922, the demands of the physics community were quite considerable, compared with supply. Ruling at Johns Hopkins had ceased with Anderson's departure; the only sources in the U.S. of new

---

[85]Carl Runge, "False Spectra from Diffraction Gratings: Part II. Theory of Lyman Ghosts," *Journal of the Optical Society of America and Review of Scientific Instruments*, 6 (1922): 429, 433.

[86]J.A. Anderson, "False Spectra from Diffraction Gratings: Part III. Periodic Errors in Ruling Machines," ibid., 437, 440-441.

[87]Anderson, "False Spectra," 442. Meggers to Anderson, 5 January 1921, WFM.

[88]Rowland, "Gratings in Theory and Practice," 536.

gratings—even those that produced ghosts—were Pasadena and the University of Chicago.[89] (Johns Hopkins kept a supply of finished gratings on hand.) The further requirement that ghosts of uncertain origin be abolished only exacerbated the dilemma.

Robert Wood, already at Hopkins, entered this situation in 1923, when he assumed responsibility for ruling operations there. Wood faced a potent critic, as had Rowland, in Lyman, still a critical user, who noted in that year that a grating received from Wood still produced ghosts.[90] Wood found a way, quite apart from theory, to visualize errors in ruling. He found that if he used an eyepiece or photographic plate to intercept the diffracted wave-front reflected from a grating at a distance short of the focal point, he could detect vertical bars of light, different from spectral lines, that would provide a fairly direct measure of errors in groove spacing.[91]

By now Wood's work was informed by the analyses of Meggers, Runge, and Anderson, and Wood directed his attention to the current value of the long periodicity of Runge's analysis, which he now found to be eighty-eight grooves. He also knew that he should consider the drive belt, and this is where he debugged the ruling engine: "a coincidence of a thick spot on the belt with the very small motor pulley at the moment when the screw is advancing the carriage, will . . . occur once in every 87 cycles of the machine." Wood changed the drive belt, and so also the periodicity, resulting in different placement of the ghosts. To abolish them, he coupled the drive pulley elastically with a second pulley, to smooth out any disturbance.[92]

Lyman received a new grating from Wood in February, 1924: "Wood thinks it free from Rowland ghosts, + as a change has been made in belt drive, free from L ghosts also." Lyman received another instrument in September, and ghosts were "said by Wood to be entirely absent with this grating." Wood had mastered the mechanics of the ruling engine, a necessary step in achieving unproblematic spectra. Lyman's Harvard colleague Frederick Saunders tested the latest grating, and Lyman noted that measurements "seem to indicate that ghosts are absent or very faint indeed."[93] Thus in October of 1924 an

---

[89]Meggers had wished to analyze gratings from the University of Chicago, checking them for the "spurious lines" exhibited by Hopkins products, but Michelson said he was unable to supply any. Meggers was skeptical: "Millikan told me that Michelson's machine was ruling most every day but I have never seen any of their perfect gratings." Meggers to A.A. Michelson, 22 August 1919; Michelson to Meggers, 28 August 1919; Meggers to Paul W. Merrill, 29 August 1919, WFM.

[90]Lyman, Notebook 1920-1926, 17 December 1923, TL.

[91]Wood, "An Experimental Study of Grating Errors and 'Ghosts,'" *Philosophical Magazine*, 48 (1924): 497-499.

[92]Ibid., 504-506.

[93]Lyman, Notebook 1920-1926, 29 February 1924, 30 September 1924, 10 October 1924, TL.

ideal was achieved, in satisfying Lyman, and falsity exorcised. Lyman later stated for the record, "It is well to emphasize the debt which spectroscopists owe to R.W. Wood for his labors in resurrecting the ruling engines at Johns Hopkins University; and for the learning and skill which he has displayed in exorcising the ghosts which haunted these historic machines."[94]

Nature was truly represented. This was, very quietly, a red-letter day in the engineering of ever-truer devices for strong dispersion of the spectrum. Only the cognoscenti, however, could appreciate this landmark. As spectroscopy began to contribute findings to the new quantum theory of the 1920s, it required perfection of some nineteenthcentury mechanical engineering. To understand the atom, Wood had to understand a drive belt. But models of the atom took centerstage in physics; the means by which physicists detected the microphysical music of the spheres was very nearly taken for granted. This was, in fact, a measure of the success of the Hopkins school, that physicists accepted the spectra from their instruments as unproblematic. Nevertheless human agency, embodied in concrete instruments, was necessary, to understand the language of nature.

---

[94]Lyman, "The Spectroscopy of the Extreme Ultra-Violet," 559.

Elements of Style

Rowland instruments formed the centerpiece of the Hopkins School of Light, creating circles of light and color in America and abroad. At the same time, they carried with them a style that affected the meaning and connotations of spectroscopic research. The style did not arise from the instrument alone, but also from the manner in which it was deployed; the mode of doing physics in Baltimore carried with it certain values. Thus the research school may well serve as an appropriate unit of analysis in which to look for traces of specific styles of doing science.[1]

There were at least four elements of the style of the School of Light that were notable. One was a continuation of Rowland's engineering approach to physical research; this approach was essentially pragmatic. Mere observation of light was not sufficient; for a complete understanding, the loose group arising from the research school simultaneously pursued control and observation, which together demonstrated their command of light. The concern was not simply knowledge, but especially the means of obtaining new knowledge. The School placed engineering in service to science, envisioning the new knowledge to be obtained with better instruments. Thus the issue of *how* to do physics was fully as important, in practice, as the issue of *what* knowledge was produced.

---

[1]The idea of style and the suitability of research schools as units of analysis for the detection of style are discussed in Mary Jo Nye, "National Styles? French and English Chemistry in the Nineteenth and Twentieth Centuries," *Isis*, 8 (1993): 30-49. The question of national style is also addressed, in the case of one research school, in Geison, *Michael Foster and the Cambridge School of Physiology*, 328-355.

A second major element of the Hopkins style was the readiness to compare terrestrial and astronomical light. Essentially, the research program that Rowland gave to his students entailed an attempt in the laboratory to produce the kinds of light that were absorbed in the solar atmosphere. Laboratory experiments, specifically the use of electric current to cause gases to glow, offered the possibility of explaining the Fraunhofer lines.

Both the inventiveness of the Hopkins school, and the readiness to analyze sunlight with reference to laboratory light, sometimes led to conceptual innovations, but these usually came from physicists or astronomers elsewhere. Conceptual change was not a desideratum of the Hopkins school; rather, the philosophical tone of the school was conservative. Rowland expected to find that the Earth and Sun were much the same: "Indeed, were the whole earth heated to the temperature of the sun, its spectrum would probably resemble that of the sun very closely."[2] Klaus Hentschel states that the "assumption of the precise coincidence of solar and terrestrial spectral lines" had been a "mainstay" of spectroscopy in the late nineteenth century.[3] This tenet, held by Rowland, was a useful working hypothesis, but a common result of his students' research showed that it was not precisely correct. The wavelengths of laboratory lines might differ slightly from the wavelengths of solar lines, and the reasons for the difference might be sought in differing pressures, temperatures, magnetic fields, or other factors. The research school could modify the assumptions under which it worked—evidence of vigor and flexibility.

There was a conservative approach evident in Rowland's general desire to maintain the principles of Kirchhoff and Bunsen, and it was also apparent in other aspects of Hopkins physics. The experimental approach, while potentially revolutionary in its implications, was not necessarily revolutionary in itself. Ames asserted, "Hypotheses rise and fall; the facts of experiment remain."[4] In Ames's view, experimental physics could be essentially a conservative enterprise, even though the facts of experiment might require conceptual accommodation or rationalization.

The conservative cast of Hopkins physics was not, by any means, necessarily political. Rather, it reflected a reluctance to alter fundamental scientific ideas, and a desire to preserve the conceptual structure of physics as it was, if possible. One of the luminaries who brought old-world physics to Baltimore was William Thomson, who

---

[2]"Report of Progress in Spectrum Work," 523.
[3]Hentschel, "The Discovery of the Redshift of Solar Fraunhofer Lines by Rowland and Jewell in Baltimore around 1890," 257.
[4]Ames, *The Constitution of Matter* (Boston: Houghton Mifflin, 1913), vi.

lectured on light and other aspects of physics at Johns Hopkins in 1884.[5] He pursued an explanation of "the dynamics of an elastic solid, especially with reference to the wave theory of light."[6] The students pooled their money and purchased a concave grating to present to Thomson.[7] Although in the 1870s Rowland had already begun to favor James Clerk Maxwell's electromagnetic theory of light in his pedagogy,[8] as late as 1888 he continued to teach that physicists could also understand light as a wave in an elastic solid,[9] as did Maxwell himself with Thomson's approach. In his lectures, Rowland still gave the two sets of equations for light, one describing light as an electromagnetic wave, according to Maxwell's theory, and the other describing light as a wave in an elastic-solid ether, according to Thomson's theory.[10] Thus, it is evident that Rowland took pains to conserve ideas, rather than overturn them.

A fourth distinctive aspect of the style of the School of Light was an internationalist awareness. As an American physicist, Rowland labored with a sense of inferiority with respect to Europe. He wrote that "no physicist of the first class has ever existed in this country,"[11] but he had studied with Helmholtz in Berlin, and received recognition for his work from Maxwell in Scotland. This training and recognition brought him into an international community of physics, of which he remained a part. The internationalist ethos, which lingered among some members of the school, may have ultimately derived from the century-old legacy of the Enlightenment. Lorraine Daston writes that the "eighteenth-century Republic of Letters created the values of universalism and the international network of communication and collaboration in the sciences."[12] In Rowland's assessment, American physics was in a state of poverty.[13] Thus, his decision to become involved in an international community was at least partly pragmatic.

An internationalist ideal took particularly strong hold in the Western scientific community around 1900, according to Elisabeth

[5]William Thomson's 1884 lectures, and accompanying historical commentaries, appear in Robert Kargon and Peter Achinstein, eds., *Kelvin's Baltimore Lectures and Modern Theoretical Physics* (Cambridge, MA: The MIT Press, 1987). Contrary to the title, Thomson was not yet Lord Kelvin at the time of the lectures. He was only elevated to the peerage in January 1892. Smith and Wise, *Energy and Empire*, 799-801.

[6]William Thomson, in *Kelvin's Baltimore Lectures*, 167.

[7]Crew, 10 October 1884, Diary, HC.

[8]See Buchwald, *From Maxwell to Microphysics*, 77.

[9]Ames, Notebook - "Rowland's Lectures on Light" (1888), 3, 13, JSA.

[10]Ibid., 47.

[11]"A Plea for Pure Science," 610.

[12]Lorraine Daston, "The Ideal and Reality of the Republic of Letters in the Enlightenment," *Science in Context* 4, 2 (1991): 369.

[13]"A Plea for Pure Science," 613.

Crawford.[14] This intellectual climate may have been propitious for the Hopkins school, as well as for the importation of European physics to the United States. In this era, scientists of various nations pursued

> an active internationalism that put scientists in the vanguard of those who worked for the betterment of the human condition, whether it be material or intellectual. . . . Because science was universal and constituted a common language, international scientific organizations, it was felt, could become models for international associations generally and even help usher in world government.[15]

The scientific community fostered not only internationalism, but also a universalist ethos that was antithetical to particularism. The universalist ethos held "that the acceptance or rejection of knowledge claims is totally independent of the personal attributes of those who make them, their nationality, race, religion, or social class."[16]

Historians of science may well reject or hold strong reservations about the validity of this form of universalism. Any historical evidence of national or local styles in science tends to suggest that, while some scientific truths may be universal, they may be expressed in different ways, depending upon the cultural setting. However, Crawford concludes that, even if the idea of universalism must be viewed with caution, "one can not disregard that in science the ease of transferability of content and methods from one language and culture to another has favored the ethos of universalism more than in other social activities." She does caution that the universalist ethos was not identical with an internationalist ethos.[17]

Crawford is not alone in detecting an undercurrent of internationalism in science at the close of the nineteenth century. J.T. Merz observed much earlier that the second half of the nineteenth century brought "greatly increased intercourse" within the international scientific community. He wrote, "This intercourse has reacted on the domain of thought, and produced that exchange of ideas which promotes more rapid progress." Perhaps ironically, Merz wrote that "the work of bringing about an international exchange of ideas has been very characteristically divided" between England, Germany, and France. Great Britain, maintaining extensive commerce, easily furnished "the modern facilities of intercourse and exchange." The scientific community of Germany took on "the complete recording,

---

[14]Elisabeth Crawford, "The Universe of International Science, 1880-1939," in *Solomon's House Revisited: The Organization and Institutionalization of Science*, ed. Tore Frängsmyr (Canton, MA: Science History Publications, 1990), 254.

[15]Ibid.

[16]Ibid., 253.

[17]Ibid., 254-255.

registering, and analysing of the scientific labours of the whole world." The French provided the first beginnings of a general and international system of units and measurements, which, like the common Latin tongue in former centuries, or like the universal languages of algebra or of music, enables us to express the results of scientific research in formulae intelligible everywhere and at all times.

Merz concluded that these three approaches to internationalization were so successful that in the late nineteenth century they began "to destroy the clearly marked differences of national thought." That era saw "everywhere in the domain of science the dying out of national restrictions."[18]

Merz, a European, ventured no generalizations about American contributions to the internationalization of science. Indeed, Rowland's own statements about the poverty of American science in his era[19] might be taken as evidence that there were few contributions. Crawford posits a mode of internationalization, however, that occurred in many countries, and that may apply very well to the particular case of the School of Light. In Crawford's analysis, a "transnational laboratory culture" provided "the underpinnings of practical international science," both before and after World War I.[20] By the 1930s, this culture demanded "more extensive—and expensive—experimental facilities," sometimes housing "novel and rare instruments." But during the heyday of the Hopkins school, prior to the beginnings of Big Science, the cost of entering the transnational laboratory culture was lower.[21] In the science of spectroscopy, it was often equal to the price of a diffraction-grating spectrograph. Thus the case of the Hopkins school suggests that the distinctively American style of promoting internationalism may have resided in the promotion of a transnational laboratory culture.

First among the traits of the School of Light was *inventiveness*, an emphasis on apparatus that was characteristic of American physics at the time. Among articles on spectroscopy in *The Physical Review*, discussions of laboratory methods and apparatus in the 1890s were more extensive than later on. From a survey of experimental reports, Charles Bazerman found that the earlier period "leaves a visual impression of detailed apparatus drawings and extensive tables of raw experimental data, while a scan of the journal of 1980 leaves a visual impression of extensive equations and schematized graphs." Through 1920, articles included many illustrations of apparatus, but such illus-

---

[18]John Theodore Merz, *A History of European Thought in the Nineteenth Century*, Vol. 1 (New York: Dover Publications, 1975; four vols. first published 1904-1912), 303-305.

[19]E.g., "A Plea for Pure Science," 594, 597, 610, 613.

[20]Crawford, "The Universe of International Science," 263.

[21]Ibid.

trations become less common after that. Bazerman found further, "Methods and apparatus sections have been generally decreasing in their proportional share of each paper."[22]

Although *The Physical Review* was not the most important vehicle for the Hopkins school of light, it is apparent from Bazerman's survey that American spectroscopy in the heyday of the Hopkins school was more instrument-oriented than it is today. Bazerman found that, over the century, articles increasingly discussed abstractions rather than objects; later writings were "centred less on concrete descriptions and more on topics of theoretical significance."[23] Because Bazerman did not limit his survey to physicists of the Hopkins school, his work suggests that an instrument-oriented approach was common; it might even represent one element of a national style in the era 1893-1920.

Clearly, among the various apparatus of the School of Light, the ruling engines held primacy. Rowland's invention of a vastly improved machine for making diffraction instruments was the *sine qua non* of the school. Perhaps to preserve the privileged position of Hopkins as an instrument provider, the engineering physicist was in fact secretive about the means of production. But instrument production was not forgotten in later periods. Instrumental know-how, while less prominent, was preserved. Later physicists at Hopkins, at other institutions, and eventually at some commercial firms, made significant improvements on Rowland's techniques. Twentieth-century innovations in ruling achieved at various locations included: the use of specially-shaped diamonds to control the exact form of grooves, which would thus exercise finer control of the spectrum; the use of improved materials and methods in engine construction; the use of interferometric control to monitor the exact motion of the ruling diamond; improved control of lubrication; the substitution of aluminum-coated glass for speculum metal as the grating material; and the use of chemical replication procedures, employing a ruled grating as a master cast. A Hopkins physicist also reported in 1960 that "gratings of high precision are no longer extremely rare as they were at the beginning of this century."[24] During the century from 1884 to 1984, according to one author, the typical worldwide production rate for good diffraction gratings of whatever size or kind had risen from ten per year to some 100,000 per year. Among the makers of ruling engines, he writes, "nearly all of them have followed [Rowland's] basic design philosophy."[25] (See Figure 1.)

---

[22]Bazerman, "Modern Evolution of the Experimental Report in Physics," 179-180, 183.

[23]Ibid., 176-177.

[24]John Strong, "The Johns Hopkins University and Diffraction Gratings," *Journal of the Optical Society of America*, 50 (1960): 1148.

[25]Erwin G. Loewen, "Current and Future Grating Technology," *Vistas in Astronomy*, 29 (1986): 223-224.

In addition to ruling, the invention called chemical replication made possible much of the eventual proliferation of affordable gratings. The technique involved pouring a liquid compound onto a ruled master grating, allowing the compound to harden, and then using the hardened compound as a grating in its own right. The method was simple and inexpensive, but the replicas were inevitably less perfect than the masters.[26] Still more recently, in the 1960s, physicists began to employ lasers as the means of forming grooves in a photosensitive material, thus forming a diffraction grating by the action of light itself.[27]

In regard to the ruling engines at the Hopkins School of Light, the most important stewards after 1901 were principally John Anderson, Robert Wood, and John Strong. For a while, Lewis Jewell operated Rowland's machines. Then, during the Baltimore fire of 1904, the machines suffered water damage. Anderson, a 1907 Ph.D., began teaching at Hopkins in 1908. Over a matter of years, he "thoroughly rebuilt" one of the engines, "and then ruled a substantial number of gratings with higher resolving power, less scattered light, and weaker 'ghost' intensities than any produced before." He also pursued the problem of making grating replicas, and studied the effect of groove shape on spectra produced by gratings. By 1912, Anderson began advising Hale on the creation of a ruling engine for Mount Wilson Observatory. In 1914, one of Anderson's titles at Johns Hopkins was Director of the Astronomical Observatory.[28]

After Anderson's departure, the ruling engines languished until 1923, when Robert Wood assumed responsibility. From about 1930, he had the abiding assistance of Wilbur H. Perry,[29] as Rowland had enjoyed the assistance of Schneider. In addition to Wood's role in eliminating Lyman ghosts, he was likely the first steward to make the operation financially self-supporting.[30] With a collaborator, Wood experimented with etched rulings on glass, for ultraviolet spectroscopy, and echelette rulings on gilded copper for infrared studies.[31] A later expert defined echelette instruments as "coarse gratings of controlled groove form designed primarily for infrared use."[32] Wood began to make and analyze these instruments as a means of exploring the

---

[26]On replicas, see C. Candler, *Modern Interferometers* (Glasgow: Hilger & Watts Ltd., 1951), 364-366; George R. Harrison, "The Production of Diffraction Gratings," 426; and M.C. Hutley, *Diffraction Gratings* (New York: Academic Press, 1982), 8-9, 125-126.

[27]Hutley, 10-11.

[28]Harrison, "The Production of Diffraction Gratings," 414. Bowen, "John August Anderson," 3-4. Strong, "The Johns Hopkins University and Diffraction Gratings," 1150.

[29]Strong, "The Johns Hopkins University and Diffraction Gratings," 1151.

[30]Harrison, "The Production of Diffraction Gratings," 414.

[31]Strong, "The Johns Hopkins University and Diffraction Gratings," 1151.

[32]Harrison, "The Production of Diffraction Gratings," 418.

effects of groove shape on the spectra produced by gratings. He wrote, "It occurred to me that a promising method of attack would be to manufacture gratings with grooves of such large size as to make the determination of their exact form, width, &c. [sic] a matter of certainty, and then investigate the energy distribution by means of the long heat-waves." Wood coined the name "echelette" to distinguish it from Michelson's "echelon" instrument.[33] The echelon was an extremely coarse grating, consisting of the edges of a series of overlapping plates, staggered to form steps about one millimeter wide. It was useful for resolving small portions of the spectrum.[34] The echelette was intermediate between a standard grating and an echelon in terms of the fineness of the intervals. Wood found that the echelette gratings could be replicated well, and the replicas mounted on glass and plated with gold.[35] Wood's studies of groove shape also held implications for creating brighter spectra in astronomical spectroscopy, as will emerge in Chapter 8.

Inventiveness maintained the ruling operation in greater vigor than simple preservation would have. Another innovation by Wood arose from his discussions and collaborations with Lyman at Harvard; Wood once again provided service to the Cambridge physicist. In Lyman's use of Hopkins instruments, there was a problem in "the oxidation or corrosion of the polished metal surface by active vapours from the discharge-tube." Thus Wood made some glass gratings as a possible alternative. At Harvard, Lyman tested one with a ruled surface of five by eight centimeters and a groove density of 15,000 per inch. Wood and Lyman together concluded that "there can be no doubt whatever that the glass grating produces a much more intense spectrum on the short wave-length side of 500 A.U. than that obtained from speculum." Therefore the new type might "prove very useful in vacuum spectroscopy."[36]

In the 1930s, even while Wood continued to oversee ruling at Hopkins, John Strong at the California Institute of Technology began to make brighter instruments possible by the deposition of aluminum films on glass. In Baltimore, Wood learned of Strong's approach and adopted it.[37] He also traveled to Caltech and together with Strong

---

[33]Wood, "The Echelette Grating for the Infra-Red," *Philosophical Magazine*, 20 (1910): 770.

[34]*Dictionary of Physics*, ed. H.J. Gray (London: Longmans, Green and Co., 1958), *s.v.* "Echelon grating."

[35]Wood, "The Echelette Grating," 778.

[36]Wood and Lyman, "Improved Grating for Vacuum Spectrographs," *Philosophical Magazine*, 2 (1926): 310, 312.

[37]Strong, "The Johns Hopkins University and Diffraction Gratings," 1151. Strong interview by David H. DeVorkin, 20 April 1984, Space Astronomy Oral History Project, National Air and Space Museum, Smithsonian Institution, Washington, DC.

studied anomalies in grating spectra. Strong afterwards worked at various institutions; in the late 1940s association with Wood and August Pfund at Hopkins led him to Baltimore. Strong brought equipment from Caltech, and also began to supervise the Hopkins ruling operation with Wood. Eventually, Strong assumed sole responsibility.[38] At that time Pfund (Johns Hopkins Ph.D., 1906) was head of the department and Wood was "ungovernable," in Strong's view. Pfund simply said: "Strong, do something about gratings."[39] (This did not imply enmity between Strong and Wood. When once asked for his recollections of Wood, Strong replied, "Oh my, you'll be here till evening, or all night as well.")[40]

Strong was admonished to maintain the position of Johns Hopkins as the leading academic supplier of gratings. One of his students suggested an unusual approach to ruling, in which groove spacing alternated; the idea was stillborn, however, as a grating of this type produced ghosts abundantly. For more conventional gratings, Strong himself devised a new way of forming an accurate screw, and Wilbur Perry produced a ghost-free grating on the very first run of a new ruling engine incorporating this screw. (By the 1980s, interferometric control of the motions of ruling engines had obviated the need for precision screws.)[41]

The instrument that attained the greatest "reach" was supplied to the federal government. For the Naval Research Laboratory, Hopkins supplied a concave grating in 1946, and the Navy placed this instrument in a Rowland mounting, which it placed, in turn, in the nose of a captured German V-2 rocket. The rocket carried the spectrograph above much of the atmosphere, in order to photograph ultraviolet portions of the solar spectrum that were not detectable at ground level because of atmospheric absorption. The first flight was unsuccessful because, as one rocket scientist remembered, "the impact of the reentering V-2 resulted in its total disappearance at the bottom of a huge crater. Nothing was recovered, in spite of much digging." The Navy tried again, however. After a subsequent rocket launch, "the V-2 was blown apart" during the descent phase. However, all was not lost, because "the tail fin with the instrument broke off and fell quite gently, and the spectrograph was recovered intact." The shortest wavelength photographed was 2400 angstroms, a moderate success. The team had ideally hoped to reached further into the

---

[38]Strong interview, 16-17, 32-35. Strong, "The Johns Hopkins University and Diffraction Gratings," 1152.

[39]Strong interview, 34-35.

[40]Ibid., 62.

[41]Ibid., 35-36, 65.

ultraviolet, to a wavelength of 2100 angstroms, but their actual attainment amounted to a success for the School of Light. A concave grating, made under one of Rowland's successors had visited the edges of the atmosphere. Over the course of two years, the Navy conducted about ten launches of V-2 rockets carrying diffraction spectrographs.[42]

To distribute gratings was, for the School of Light, to disseminate a way of doing physics, but Strong produced physicists as well as spectroscopic instruments. During his tenure in Baltimore, he taught more than two dozen Ph.D. students, thus combining instrument-making, research and pedagogy as Rowland and Wood had done, and continuing the School of Light. Like Rowland, Strong came to spectroscopy in mid-career; it had not been his interest while a student. Strong recalled, "My teaching of classes was without benefit of ever having had a course as a student in optics or spectroscopy. I don't recommend that—but, it does give a fresh point of view." He continued to teach optics for years. In the 1950s, Strong also became director of the Laboratory of Astrophysics and Physical Meteorology. He kept a guidebook to experimental physics in print for 40 years, through 35 printings, and felt that its principal usefulness was that "i̇ ̇erved young fellows in determining whether or not experimental physics was to their talents, tastes and personality traits sufficiently to promise a rewarding career in it."[43]

In general, Rowland instruments were dedicated to the practice and teaching of research physics, and the allocation of instruments entailed allocation of the capacity to pursue state-of-the-art spectroscopy. Adhering to the intentions of the 1880s, producers of Hopkins instruments tended to reserve their products for the exclusive use of research physicists. The eventual dissemination of the instruments beyond the domain of pure research, in the 1930s, relied on the mediation of a commercial firm that repackaged concave gratings into marketable spectrographs.[44] This hardly led to a demise of disinterested research; the commercial availability of diffraction instruments led to their more widespread deployment in chemical research, Robert Wood's original interest at Hopkins in the 1890s.

Baird Associates in Cambridge, Massachusetts made available "the first commercial spectrograph suitable for the qualitative and quantitative analysis of iron, steel and minerals, and it demonstrated

---

[42]Richard Tousey, "Solar Spectroscopy from Rowland to SOT," *Vistas in Astronomy*, 29 (1986): 178, 181.

[43]Strong interview, 39, 48, 63, 65.

[44]Davis Baird, "Baird Associates's Commercial Three-Meter Spectrograph and the Transformation of Analytical Chemistry," *Rittenhouse*, 5 (1991): 65-80.

the advantages of a grating spectrograph over a prism spectrograph for chemical analysis."[45] Walter Baird, one of the co-founders of the company in 1936, had received his Ph.D. in electrical engineering from Johns Hopkins, where he studied physics under Wood, who "taught him to bring physical principles and techniques to the construction of new instruments." Baird's acquaintance with Wood likely gave him better access to the source of supply of Rowland instruments, and the concave diffraction grating was indeed the critical component of the Baird spectrograph.[46]

Wood had declined to supply instruments to another firm, Bausch & Lomb, because he viewed their enterprise as exclusively commercial. But he knew Baird, and was "intrigued by Baird's ideas for a self-contained grating spectrograph suitable for chemical analysis." Baird also knew Wood's assistant, Wilbur Perry. Therefore, Wood began to supply Baird with the highly reflective gratings made possible by aluminum surfacing techniques. Eventually, in 1948, Baird Associates built their own ruling engine.[47]

Baird essentially popularized the Rowland grating by repackaging it in a ready-to-use spectrograph, so that even those unversed in the ways of the physical laboratory could benefit from using it. The "package" amounted to "a stable, self contained unit which could be easily operated by people who lacked expertise in spectroscopy." Baird sold the unit to government agencies, universities, and corporations, and continued to do so into the 1960s. In the 1940s, Baird and Dow Chemical devised a "direct reading spectrograph," which "automated the interpretation of spectral lines, allowing analysts to determine the concentration of selected elements in minutes where previous spectral analyses took hours." The Baird spectrograph led the way in making Hopkins-style high-dispersion spectroscopy available to anyone with a practical use for it, and the means to afford it. Rowland instruments were no longer solely a means of inculcating virtue into apprentice physicists and ordering the unruly light of the sun and the chemical elements. They had also become an extremely practical means of obtaining subtle knowledge of the composition of complex substances, and in this respect pure research eventuated in pragmatic results.[48] The Baird spectrograph took Rowland gratings beyond the confines of research schools and research physics, but simultaneously obscured the agency of the school, by placing the gratings inside a larger package.

---

[45]Ibid., 65.
[46]Ibid., 65, 67, 75.
[47]Ibid., 67.
[48]Ibid., 67, 70, 73, 75.

Although the principal work of the school generally concerned solar and laboratory spectra, in the 1880s Rowland already thought about spectroscopy of the stars, and extending the *celestial-terrestrial combination* that was a second trait of the School of Light. He envisioned the deployment of concave gratings of short focus, creating a relatively small spectrum, so that the relatively faint spectra of stars would be concentrated in a small space, and might thus be bright enough to capture photographically. Spectra from short-focus instruments had to form on a surface with significant curvature, however, and Rowland therefore envisioned the use of paper or celluloid photographic negatives, which could maintain curvature much more readily than glass plates.[49] But he never implemented his ideas.

Henry Crew carried out the first significant attempt to extend the celestial-terrestrial connection and pursue stellar spectroscopy with a Rowland instrument. He had moved to Lick Observatory in California in 1891, with hopes of applying the instrument in general astronomy, rather than solely the solar observations conducted in Baltimore. Although Crew was able to promote and carry out a pilot project with the concave grating at Lick, he captured only a part of the starlight coming from the thirty-six-inch refractor telescope at Lick, and the resulting spectra were too faint to be useful; the project was discontinued. Crew took a permanent position at Northwestern University, partly at the urging of George Ellery Hale in Chicago, who said he wanted a fellow spectroscopist "within easy reach."[50] In the 1890s, Frank Wadsworth studied possible mountings or configurations that might actually make concave instruments useful in astronomical spectroscopy.[51]

In 1902, Hale used a Wadsworth mounting, which greatly reduced the natural astigmatism of the concave grating. When he was still experimenting with the apparatus in order to take stellar spectra, a wooden telescope housing on the grounds of Yerkes Observatory caught fire, and destroyed the spectrograph. A year later, he was building a new horizontal telescope as a replacement, and waiting for a new Rowland grating from Baltimore. In 1905, he moved this telescope to Mount Wilson in California, but for a pioneering diffraction study of Arcturus, he used a plane grating, because he could not acquire a concave grating soon enough. Still, the apparatus yielded spectra of great dispersion and resolution. Donald Osterbrock observes, "Concave gratings never did catch on for stellar spectroscopy,"

---

[49]Ames, "The Concave Grating in Theory and Practice," 377.
[50]Donald E. Osterbrock, "Failure and Success: Two Early Experiments with Concave Gratings in Stellar Spectroscopy," *Journal for the History of Astronomy*, 17 (1986): 119-123. Hale to Crew, 14 July 1892, HCM.
[51]F.L.O. Wadsworth, "The Modern Spectroscope. XIV. Fixed Arm Concave Grating Spectroscopes," *ApJ*, 2 (1895): 370.

although further attempts were made. One problem was that concave gratings possessed astigmatism that tended to "widen and significantly weaken the spectrum." Osterbrock adds, "However, for a few applications, particularly ultraviolet spectroscopy from rockets and space vehicles, [concave gratings] are unmatched."[52] Eventually, plane gratings did become useful for astronomical spectroscopy from ground level, after improvements at Johns Hopkins in the 1930s made possible much brighter spectra. (See Chapter 8.)

Although Crew returned to physics proper at Northwestern, he kept an interest in celestial matters and maintained an ongoing dialogue with Hale. He developed methods of approaching extreme celestial conditions in the laboratory, and Hale sometimes attempted to reproduce these techniques, usually involving the electric arc and spark, the closest approaches to celestial heat. At Yerkes, in 1901, Hale duplicated some of the apparatus that Crew had developed at Northwestern. Hale also constructed a spark device in a steel chamber that would withstand a pressure of several hundred atmospheres, to test effects on spectra.[53] Five years later, at Mount Wilson, Hale received an arc-producing device from Crew. Hale believed that results achieved with the device helped in interpreting sunspot spectra.[54] The continuing esteem of the astronomical community toward Crew was evident in the 1904 offer of the directorship of Allegheny Observatory in Pennsylvania, which Crew declined.[55]

Laboratory reproductions of celestial conditions remained intriguing to Hale and thus to the research community that gradually formed at Mount Wilson Observatory. An enduring interest there concerned the nature of vortices at the surface of the sun, and John Anderson, when visiting from Hopkins, made pertinent laboratory studies. Hale later told him of "our recent vortex experiments, which confirm the work you did here. The secondary smoke vortices are very beautiful and closely resemble some of the *Ha* flocculi."[56] Anderson's published *oeuvre* does not include any vortex studies; in fact, for some six years after encountering Hale, Anderson did not publish at all. The studies to which Hale referred apparently resembled similar efforts by Crew and Wood, to re-create in the laboratory some aspects of solar behavior or appearance.[57]

In Anderson's research, one recurring subject was the Stark effect, under both terrestrial and celestial conditions. It was complementary

---

[52]Osterbrock, "Failure and Success," 124-128.
[53]Hale to Crew, 29 March, 8 May 1901, HCM.
[54]Hale to Crew, 29 May, 11 June, 9 October 1906, HCM.
[55]John A. Brashear to Crew, 26 November 1904, HCM.
[56]Hale to Anderson, 19 June 1915, GEHM.
[57]Anderson to Hale, 4 July 1915, GEHM.

to Hale's earlier work on the Zeeman effect in the sun, his most important single finding. In conjunction with laboratory studies, Anderson may well have hoped to detect a vertical electric field, like that existing in the terrestrial atmosphere, in the atmosphere of the sun. The electric Stark effect was essentially an analogue of the magnetic Zeeman effect. However, Anderson's experiments on the Stark effect were cut short by the military needs imposed by the First World War.[58]

Anderson continued to pursue laboratory studies of various kinds, with the hope that it might be possible to "imitate successfully the spectra given by the sun and stars." To do so, he relied on extremely high voltages, in order to obtain very high temperatures and pressures. Using a large condenser, Anderson created a potential of 26,000 volts, to drive a current through a wire that was typically 0.127 millimeters in diameter or less. As a result, "the wire blew up like an ordinary fuse with a blinding flash of light." The temperature of the explosion, occurring in 10^{-5} seconds, briefly reached 20,000 degrees, and Anderson was able to take a continuous spectrum, crossed by absorption lines, from the flash. The grating was "a 10 cm (4 in.) Rowland, known as the 'Kenwood grating.'" The creation of an absorption spectrum was especially useful, because most high-temperature laboratory sources gave only emission lines, while spectra of the sun and stars revealed a multitude of absorption lines.[59] Rowland earlier performed this type of experiment, using an induction coil, together with an alternating Siemens dynamo and several Leyden jars, to send a pulse of current through an iron wire 1/16 inch in diameter.)[60]

In one instance, explosion of the wire briefly created a cylinder of luminous gas that expanded at ten times the velocity of sound, reaching twenty-two millimeters in diameter. In the ephemeral light of this gas, Anderson sought to observe sun-like conditions, but found that his source actually possessed one hundred times more surface brightness than did the sun. Anderson discussed his results in terms of a meteoric explanation of solar heat, after Kelvin. According to this understanding, during a meteor's descent into the solar atmosphere, as during the vaporization of a wire, "*a very large quantity of energy is thrown into a small amount of matter in a short space of time.*" Anderson allowed that differences existed: in falling into the sun, his wire would give up 80,000 calories of heat, rather than merely the thirty given up in the laboratory. But still he had managed to obtain absorption lines. Believing iron to be "perhaps the most important element from the

---

[58]Bowen, "John August Anderson," 5.
[59]Anderson, "The Spectrum of Electrically Exploded Wires," 37, 38, 40, 42.
[60]Ames, "The Concave Grating in Theory and Practice," 376.

standpoint of the astrophysicist"—because of the great many iron lines that appeared in the solar spectrum—Anderson used iron as the material for the wires. (The many iron lines suggested that iron was the most abundant element in the sun, but it was only after 1920 that the relative scarcity of iron in the sun was appreciated.) Sounding a theme later characteristic of twentieth-century nuclear physics, Anderson believed that more answers to experimental questions would come with greater energies.[61]

Anderson remained in charge of the Pasadena production of diffraction gratings until 1928. By then, his skill qualified him to oversee the production of a still more celebrated instrument, the 200-inch reflecting telescope that entered operation on Mount Palomar twenty years later and remained the largest telescope in the world for decades. After Hale received funding for the project in 1928, Anderson also planned for the laboratory and machine shop that the new observatory would need. The most significant technical problem for Palomar was casting the glass disc for the main mirror, a problem given to a contractor, but Anderson also analyzed the potential effects of turbulence in the terrestrial atmosphere on the fidelity of images. The atmosphere, Anderson reasoned, made a star image appear ten to twenty times larger than it would otherwise be. It also reduced the brightness to one percent, or less, of the brightness that would be observed in the absence of an atmosphere. In this sense, the air in the terrestrial sky changed the image of a star into "a boiling patch of light." As telescopes became larger, and brought ever-smaller images into view, the deleterious effects of atmospheric turbulence became noticeable.[62]

As Anderson continued to guide the Palomar project, he rapidly found that he was in charge of a cultural phenomenon. Workmen poured the glass for the enormous reflecting disk at Corning, New York in 1934. Anderson then had an offer from a trucking firm willing to haul the large disk across the continent with great fanfare. An entrepreneur there wrote, "Recently, Lowell Thomas rated the making of this 200" lens the greatest item of interest to the civilized world in the last twenty-five years, not even excluding the great war." The trucker proposed to capitalize on the sensation:

> My idea would be to make it a point to stop over night [sic] at as many cities and larger towns as possible while enroute. The newspapers and news reels would seize upon this undertaking and would no doubt play it up to a great extent.[63]

---

[61]Anderson, "Spectrum of Electrically Exploded Wires," 38, 39, 44, 45, 48. Emphasis in original.
[62]Bowen, "Anderson," *BMNAS*, 9-12. Anderson, "Larger Telescopes" (1929), in Anderson Correspondence, GEHM.
[63]Glen M. Wiley to Anderson, 10 May 1934, in Anderson Correspondence, GEHM.

Although the mirror was eventually transported west by train, thousands of people along the way came to see it pass.[64]

Anderson and others realized that the largest telescope in the world would only be as good, in use, as the auxiliary instrumentation accompanying it. Ira Bowen later observed that, "on the instrumental side of the 200-inch project, the biggest single contributor was John Anderson."[65] (Bowen became director of Mount Wilson and Palomar Observatories in 1948, when astronomers began preliminary observational tests with the 200-inch telescope. The telescope entered regular service in 1950.)[66] Anderson supervised the program of developing new spectrographs for the planned observatory, and some of these were tested on the existing 100-inch telescope on Mount Wilson. Bowen stated that these spectrographs "made possible many new fields of spectroscopic study including the observations which led to the concept of the expanding universe."[67] It was the creation of instruments that underpinned Anderson's odyssey from terrestrial physics to celestial studies.

While the instruments created by various generations of the Hopkins school yielded observations and data that sometimes held revolutionary implications, the physicists of the School of Light were little inclined to jettison their well-established heritage in physics, keeping the *conservative tone* that was a third trait of the school. Some of the elements of physics that they conserved were very material, including the concave gratings. Rowland himself had given a caution:

> In learning the use of any new instrument the student should first study the instrument very carefully with the eye alone and find out all he can about its adjustments before touching it. This practice will not only cultivate the student's powers of observation, but may also save valuable instruments from receiving serious damage through misuse.[68]

Although Fraunhofer's ruling art in the early nineteenth century did not survive him, the very existence of a research school in spectroscopy at Johns Hopkins tended to conserve the sometimes tacit skills necessary to make and use spectroscopic instruments.

The spectroscopists were also little inclined to modify the concepts that served as intellectual or heuristic instruments by which to organize the facts of nature that they amassed. Rather than change their organizing concepts in order to accommodate the plethora of facts, they

---

[64]Wright, *Explorer of the Universe*, 420-421.
[65]Bowen, "Anderson," *BMNAS*, 12.
[66]Horace W. Babcock, "Ira Sprague Bowen," *BMNAS*, 53 (1982): 98-100.
[67]Bowen, "Anderson," *BMNAS*, 11.
[68]"Guidelines for Laboratory Work," 1, Box 39, HAR.

mainly preferred to conserve those concepts and assimilate facts to the conceptual structures that already existed. Ames held a gradualist view of progress in physics: "a continuous increase in knowledge, not an overthrow of past achievements, is to be expected." Progress meant continued experiment. He wrote, "The value of a good hypothesis consists, to a large degree, in the stimulation it gives to new experiments; for, whether the hypothesis is verified or not, the facts as found by the experiments remain, they are permanent and of ever-enduring usefulness."[69] Hopkins-school physicists clung to the guiding principles of Kirchhoff and Bunsen, even if, as Hentschel has pointed out, measurement did not always confirm the expectation of an exact coincidence between solar and laboratory lines.[70] In 1920, Anderson continued to hold the nineteenth-century view that solar heat was a consequence of meteoric infall; it informed his laboratory experiments.

To conserve and maintain an awareness of past achievements in physics, primarily those in Europe, Ames supervised the publication of a reprint series of twelve volumes of classical memoirs in physics.[71] He placed Crew in charge of selecting and editing the memoirs concerned with light and heat.[72] Crew characterized older writings on the wave theory of light as kinematical, in contrast to the dynamical mode of explanation employed in the physics of his own time. Crew stated that dynamical analysis called for consideration of the structure of the luminous body and also of the medium carrying light.[73]

In bringing together classical memoirs in physics, Crew's command of language was undoubtedly helpful. He had studied Latin and Greek during secondary school, and also as an undergraduate at Princeton, but had always been more enthusiastic about science than language. Later, he collaborated with Alfonso de Salvio on a new translation, published in 1914, of Galileo's *Dialogues Concerning Two New Sciences*. Crew regarded Galileo as "the founder of modern physics." (Hale was more concerned with Galileo's material culture: in 1923 in Italy, with the help of Giorgio Abetti, he brought some of Galileo's original telescopes from Florence to Arcetri, and observed the heavens, to "'see with his eyes.'") Crew also published, in 1928, a history entitled *The Rise of Modern Physics*. He was a founding member of the History of Science Society, and became president of it in 1930. In 1940, Crew

---

[69]Ames, *The Constitution of Matter*, 218-219.

[70]Hentschel, "The Discovery of the Redshift of Solar Fraunhofer Lines by Rowland and Jewell," 257-263.

[71]Crew, "Joseph S. Ames," 187.

[72]Ames to Crew, 5 February 1898, HCM.

[73]Crew, "Preface," in *The Wave Theory of Light: Memoirs by Huygens, Young and Fresnel*, ed. Crew (New York: American Book Company, 1900), v-vi. This reprint edition is in turn reprinted in *The Wave Theory of Light and Spectra*, ed. I. Bernard Cohen, along with memoirs edited by Ames.

released his translation of a medieval treatise on optics written in Latin by Maurolycus. (He promoted contemporary as well as classical physics, devoting three-and-a-half years to plans for a Hall of Science for the Century of Progress Exposition that opened in Chicago in 1933, using light from the star Arcturus to activate the opening.)[74]

Crew's desire to preserve the heritage of physics did not make him a scientific reactionary. He especially appreciated new methods of instruction, including the teaching laboratory. He believed that the "energy idea"—the principle of the conservation of energy and attendant transformation of dynamics articulated in the mid-nineteenth century—had not only unified physics but had also given physics instructors a more logical way to present their subject. Crew was an "ardent admirer" of Herbert Spencer, with his ideas of progress, and in 1904 was glad that "the engineer and the physicist are closer friends than they were twenty-five years ago."[75]

Ames displayed a more Germanic influence. Both he and Crew had received part of their training in Germany, and Ames tried to preserve knowledge of the German physics tradition. In the first years of Johns Hopkins, it was common for students to study in Europe, to deepen their knowledge. Ames recalled, "Students were going in considerable numbers to Germany, . . . and were returning preaching a new gospel."[76] In the 1880s, Ames followed in the steps of his mentor Rowland by working in Helmholtz's Berlin laboratory.[77] After the student years of Crew and Ames, it became less necessary for students to travel abroad for education in physics, but Ames retained an interest in German developments. Otto Laporte, a student of Arnold Sommerfeld, came from Munich to Washington, DC in 1924 to conduct research on spectroscopy at the National Bureau of Standards, where William Meggers was working. Meggers and others there took an interest in learning about mathematical achievements in Europe pertaining to the rationalization of spectra. (The flow of knowledge was not unidirectional. Sommerfeld had been to America, and a student later recalled that he "came back loaded with spectroscopic information." At a Munich colloquium, "instead of having a blackboard full of fancy equations and maybe curves, he had enormous plates of the calcium spectrum which he had gotten in Pasadena, I think.")[78]

---

[74]Meggers, "Henry Crew," *BMNAS*, 35-36, 38, 39, 40. Crew, *General Physics* (New York: The Macmillan Company, 1909), 2. Wright, *Explorer of the Universe*, 346. Knowlton, "Henry Crew," 169.

[75]Crew, "Recent Advances in the Teaching of Physics," 482, 484-485.

[76]Ames, "Recollections of a University Professor," 210.

[77]Kevles, "Ames, Joseph S.," *DSB*, Vol. 1 (1970), 132-133.

[78]H.R. Crane and D.M. Dennison, "Otto Laporte," *BMNAS*, 50 (1979): 274. Otto Laporte, interview by Thomas S. Kuhn, 29 January 1964, 20, Archive for History of Quantum Physics (AHQP), American Philosophical Society Library, Philadelphia.

The staff working at the Bureau of Standards included physics students from nearby Baltimore, and Ames hoped that Laporte could impart some of the knowledge he brought from Germany to Hopkins students. Laporte was summoned to Ames's office, and he later recalled,

> Professor Ames was wearing a light grey cut-away, looking for all the world like a German professor. He said to me that something had to be done about the education of the younger men at the Bureau of Standards because so many were taking their degree at Johns Hopkins, with no chances to have the right kind of education.

Because Laporte had studied with Sommerfeld, Ames asked him to give evening lectures four or five times a week, after the working day was finished at the Bureau.[79] Ames here appeared eager to represent the German tradition in physics to his latter-day American students.

Laporte saw a contrast between the styles of Ames and Wood. He stated, "They did not get along with each other; Ames was the theoretician and Wood the experimentalist. Ames was not productive and was just an organizer; Wood was terribly imaginative and productive." Laporte felt that, if the two had cooperated more, the department might have thrived more.[80] In one analysis given by Daniel Kevles, Rowland's death in 1901 "marked the beginning of a precipitous decline in the quality of the department."[81] However, as seen in Chapter 3, the department continued to produce physicists, instruments, and spectroscopic research well into the 1920s. A separate analysis by Kevles suggests that Ames "kept the physics department at Johns Hopkins alive despite persistent budgetary problems by cooperating with the National Bureau of Standards."[82] Eventually, the department lost Ames when he became provost of the university in 1926, and president in 1929. But he conserved the institution of which he was a part, while Wood preserved instrument production.

Ames also expressed a conceptual conservatism, or at least ambivalence, with respect to the theory of relativity. In 1913, he retained the concept of an ether, a medium for light waves, even though it was ruled out in the special theory of relativity.[83] After the emergence of the general theory of relativity, Ames assessed it partly because of the great publicity given to it, and also because it was "a subject which he was repeatedly called upon to explain."[84] Ames wrote, "I fear that the

[79]Laporte, interview with Kuhn, 31 January 1964, 23, AHQP.
[80]Laporte interview, 31 January 1964, 23.
[81]Kevles, *The Physicists*, 80.
[82]Kevles, "Ames, Joseph S.," 133.
[83]Ames, *The Constitution of Matter*, 27-28, 235.
[84]French, *A History of the University*, 409-410.

attention of most of us was first directed seriously to the matter by . . . articles in the newspapers." Thus he felt that it was his responsibility to explain relativity as best he could to colleagues. Ames wrote that the general theory of relativity contained "certain mathematical laws for a gravitational field, laws which reduced to Newton's form in most cases where observations are possible, but which led to different conclusions in a few cases." He also observed that the formulae had been "confirmed in a variety of ways and in a most brilliant manner."[85]

Almost as an afterthought, Ames displayed his conservatism when he expressed some reservations. He wrote, "There is not the slightest indication of a mechanism, meaning by that a picture in terms of our senses. In fact what we have learned has been to realize that our desire to use such mechanisms is futile."[86] Thus Ames acceded to the abandonment of a mechanistic approach. He wrote a year later, however,

> in using the word 'theory' of any phenomenon we ordinarily mean that we are proposing a picture or description of it in terms of more familiar or more simple facts. Now Einstein does not in any sense give us a picture of gravitation, nor does he attempt to describe it in terms of other phenomena.[87]

Thus Ames was not an enthusiast for the theory.

Twenty years earlier, Crew had suggested the kind of model that physicists of his day expected. He wrote,

> there is now widespread opinion that any physical phenomenon is "explained" only when someone has devised a dynamical model which will duplicate the phenomenon. The completeness of the explanation is to be measured by the completeness with which the model will duplicate the phenomenon.[88]

This was probably Ames's measure, too. One salient problem at the beginning of the twentieth century was the nature of the microphysical objects emitting light, presumably atoms or molecules. The complexities facing Crew's contemporaries were daunting. He wrote,

> it must never be forgotten that in all probability the vibrating atom is a structure whose motion is vastly complicated as compared with the few simple motions which the experiments of Huygens, Newton, Young, Fresnel, Maxwell, and Michelson have assigned it.[89]

But while the concepts of these eminent physicists might require modification, Crew very much held to the need for a vibrating model.

---

[85] Ames, "Einstein's Law of Gravitation," *Science*, 51 (1920): 253, 260.
[86] Ibid., 261.
[87] Ames, "Einstein's Principle of Relativity and Its Bearing Upon Physics," 2, 16.
[88] Crew, "Preface" (1900), v.
[89] Ibid., xiii-xiv.

Physicists of the School of Light tried to understand light-emitting objects on analogy with macroscopic resonators, thus pursuing mechanical models, consistent with the desiderata expressed by Ames and Crew. The former asserted that it was natural for a physicist to "attempt to devise a mechanical model, i.e., a material mechanism consisting of pulleys, levers, wheels, cords, etc., which should have properties as nearly as possible like the phenomena concerned."[90] Rowland himself had essentially suggested that vibration in molecules and atoms resembled the production of musical notes, thus implying an acoustical analogy, with musical instruments.[91] Ames used an analogy with tuning forks.[92] W.J. Humphreys conjectured that molecules could be compared with rectangular steel bars, in that their spatial dimensions might increase with increasing temperature, and thus the wavelengths of their vibrations would increase, as well.[93] Hopkins spectroscopists were loath to give up the use of mechanical models, such as the acoustic analogy, to explain spectral lines. Thus, although the use of new instruments brought a plethora of new data, these physicists were essentially conservative with respect to their organizing concepts, as well as their institution and their continuing emphasis on instrument production.

In the case of Rowland, the fourth trait of *internationalism*, participation in an international community of physicists, was essential, as evidenced by his work in Helmholtz's laboratory and the importance of Maxwell's recognition from Scotland in securing Rowland's appointment at Hopkins. And Rowland's ongoing membership in many European scientific societies[94] always brought a larger sense of participation. When Ames, Keeler, Crew, and others traveled to Germany to supplement their studies in Baltimore, they were essentially retracing the route of Rowland, when Gilman enabled him to complete his education in physics abroad.[95] Thus, as the School of Light became established, it was still often necessary for students to supplement their education abroad, and the internationalist aspect of the Hopkins style initially arose from this necessity. As Crew reported, "At the time when Ames matriculated, there was, indeed, a 'Johns Hopkins University Club' to which the sole requirement for admission was that the applicant had previously studied in some foreign

---

[90]Ames, *The Constitution of Matter*, 219-220.
[91]"The Highest Aim of the Physicist," in *PPR*, 671.
[92]Ames, *The Constitution of Matter*, 146.
[93]Humphreys, "Changes in the Wave-Frequencies," 228.
[94]See p. iii, *PPR*.
[95]Rowland's work there is described in Samuel Rezneck, "An American Physicist's Year in Europe: Henry A. Rowland, 1875-1876," *American Journal of Physics*, 30 (1962): 877.

university." Indeed, the custom of foreign study was a "veritable craze" among Americans in the 1870s and 1880s.[96]

After the School of Light began to flourish, foreign study became less necessary, as adequate resources were available in the United States. However, some Hopkins-school physicists, and also Hale, continued to maintain important contacts with scientists abroad. One of them was Robert Wood, who "always had close ties with British scientists." In 1900, he was asked to lecture in England on color photography and the imaging of sound waves. He received various honors from the English, and often published his work in the British *Philosophical Magazine* or the *Proceedings of the Royal Society*. A colleague concluded that Britain, "perhaps more than any other country, showed appreciation for Wood's genius."[97]

Another Hopkins-school physicist who maintained significant ties with Europe was William Meggers. He saw no conflict inherent in intercourse with German scientists after the First World War, when they were excluded from some western scientific associations as former belligerents. He blamed the scarcity of German-language physics publications in America on "the irregularities and unreasonable censorship of the mails."[98] In the aftermath of the war, he tried to reopen relations with physicists in Central Europe, and wrote to Heinrich Kayser, "Communication has been difficult or impossible for several years and we have little knowledge of . . . recent spectroscopic work in Germany, Austria and Russia."[99] (Meggers remained close to Kayser, who ran one of the leading spectroscopic laboratories in Germany and, like Meggers, pursued encyclopedic knowledge of the spectra of the elements.[100]

As Meggers re-opened channels, he found his own research strengthened by international contacts. He learned that J.M. Eder of Vienna might sell a collection of spectrograms and samples of rare earths and other elements. Meggers agreed to buy the collection, plus details on the history of the samples. Physicists at the Bureau of Standards had already been working on rare earths for eight months. He informed Eder, "We will be glad to continue your good work." For Meggers this was an opportune find, coming at the beginning of his career. He received fifty samples, and 162 spectrograms. When he learned that Eder actually needed relief aid, he wrote that he could

[96]Crew, "Joseph S. Ames," 185.

[97]Dieke, "Robert Williams Wood," 337.

[98]Meggers to K.W. Meissner, 8 January 1919, WFM.

[99]Meggers to Heinrich Kayser, 1 July 1919, WFM.

[100]Dörries, "Heinrich Kayser as Philologist of Physics," *Historical Studies in the Physical and Biological Sciences* 26, part 1 (1995): 4, 10, 20, 25.

only bemoan the combination of "insane militarism" with the "present starvation of the world's intellectuals."[101]

One twelve-week sojourn by Meggers in Europe included visits with prominent leaders in Oxford, Cambridge, Leiden, Bonn, Göttingen, Berlin, Vienna, Munich, Tübingen, Paris, and Florence.[102] Meggers served for a decade as head of a commission of the International Astronomical Union devoted to the recommendation of international wavelength standards. He especially hoped to include spectroscopists in formerly belligerent German-speaking countries. But his own published work revealed only a sustained research program in laboratory spectroscopy, not a program of international cooperation. In Meggers's career, internationalism was a matter of style, of the conditions under which he wished to work, as he related to Charles St. John, a laboratory and solar spectroscopist at Mount Wilson:

> I am going to tell you frankly how much we regret that the Astronomical Union is not truly international. What ultimate good can come from further bitterness and narrow-mindedness in dealing with scientific facts? . . . Further than this, great harm can result from the short-sightedness which not only restricts cooperation but suggests that the nationalities excluded from our 'international' Councils and Unions will be equally justified in forming similar 'international' Councils and Unions of their own.[103]

Meggers long remained an idealist, but his organizational activities were most successful when they grew out of particularist research. This supports Elizabeth Crawford's argument that, "[j]ust as in prewar times, the transnational laboratory culture provided the underpinnings of practical international science."[104]

Of all those associated with the School of Light, the most active in promoting an internationalist approach was Hale, a satellite of the Hopkins school and not a direct product of it. In the 1890s, Hale polled the board of editors of the *Astrophysical Journal* on the best use of symbols and nomenclature, especially with regard to spectral lines. Rowland's solar tables expressed wavelengths in units of 10^{-10} meters, with fractions stated to three decimal places. This unit was formally declared to be the *"ten millionth of a millimeter,"* with the aside that it was "Ångström's unit" and was also sometimes called the "'tenth-meter.'" French physicist Alfred Cornu expressed his regret that the unit was not labeled according to the system of prefixes furnished by the metric

---

[101]Meggers to J.M. Eder, 16 January 1920; 16 February 1920; 21 April 1920; 12 May 1920; WFM.

[102]Meggers to Alfred Fowler, 6 November 1921, WFM.

[103]Meggers to Charles E. St. John, 19 January 1920, WFM.

[104]Crawford, "The Universe of International Science," 263.

or C.G.S. system. For the German *Astronomische Nachrichten*, William Huggins and H.C. Vogel had already suggested adoption of the symbols H_a, H_β, etc., for the hydrogen spectral series known today as the Balmer series, and the *Astrophysical Journal* followed their lead. The board also agreed to print spectra with longer wavelengths toward the right, and wavelength tables with the shortest wavelengths at the top. The board expressed the hope that all astronomers, physicists, and astrophysicists would adopt the same standards.[105]

Hale did not attribute his internationalist ideals to the Enlightenment, but the survival of the Enlightenment ideal of internationalism is evident in the style of his work. (Daston writes of Enlightment serials, "The editorial policies and the readership of these learned journals aimed to be international; indeed, it was often their *raison d'être*.")[106] Hale's internationalism included a note of national pride. Although the editorial board of the *Astrophysical Journal* was international in composition, half the members were American, and Rowland's tables were completely so. Parisian spectroscopist Henri Deslandres discerned, in the initiation of the journal, American science taking a leading role, bringing Europe in train.[107]

Hale also expressed his workshop diplomacy in the creation of the International Union for Cooperation in Solar Research in 1904, and in the transformation of that organization to a general astronomical union, which began at Mount Wilson in 1910. The inception of the IUCSR took place at the St. Louis Fair of 1904, celebrating the centennial of the Louisiana Purchase. In discussions with physicists and astronomers at the Fair, as in his own work, Hale stressed solar research—on the most accessible star.[108]

The IUCSR met in various cities after that, sometimes in Europe. When it met in 1910, Hale provided the context himself—Mount Wilson. There, the union changed in a way that reflected developments on the mountain. Having started with solar telescopes and solar observations, the observatory in 1908 became a leader in general astronomy with the installation of a sixty-inch telescope. As for the IUCSR, "in 1910, its scope was extended to include the whole subject of astrophysics."[109] It also continued to set an agenda in spectroscopy. The

---

[105]"Action of the Editorial Board of the Astrophysical Journal with Regard to Standards in Astrophysics and Spectroscopy," *ApJ*, 3 (1896): 2-3. Emphasis in original. Rowland, "Preliminary Table of Solar Spectrum Wave-Lengths. I," *ApJ*, 1 (1895): 32. William Huggins to Hale, 10 November 1893, GEHM.

[106]Daston, "The Ideal and Reality of the Republic of Letters," 376.

[107]In Huggins to Hale, 6 January 1895, GEHM.

[108]Wright, *Explorer of the Universe*, 189-190.

[109]Hale to Elihu Root, 10 March 1913, in *Science in America: A Documentary History, 1900-1939*, 204.

group decided that an agreement among three sets of observations would be necessary for the establishment of certain standard wavelengths as benchmarks. In effect, the three sets were American, French, and German. The measurers reached agreement on the wavelengths of eight-five spectral lines.[110]

During the meeting at Mount Wilson, one proposal arose that Harvard astronomer Edward Pickering thought was "likely to revolutionize the work of the Union." The group voted unanimously to extend the scope of the union beyond the sun to the stars. This enlargement of the scope of the union would, Pickering wrote, "make it like the Astronomische Gesellschaft" (the major German astronomical society).[111] Although the scope of the union expanded, the formal establishment of a successor organization did not take place until after World War I.[112] Then, no longer feeling constrained to follow the protocols of the German-based International Association of Academies, Hale and others developed new forms of international association. A newly constituted International Astronomical Union began to emerge at about the same time that Hale fostered an umbrella organization, the International Research Council, intended to subsume many fields of science.[113] At a 1919 organizational meeting in Brussels, the new astronomical union laid claim to a large scope, with special committees chartered to manage specific fields of research.[114] Thus it was Hale, the solar spectroscopist who relied so heavily upon the productions of Johns Hopkins, who became the most effective advocate of international cooperation, and the new union held a very large scope both in membership and in subject matter considered.

The inventiveness that formed the basis of the Hopkins school, the research program of comparing light from the sky and from the laboratory, the conservatism with respect to guiding principles, and the internationalist approach, were not entirely separable components of the Hopkins style. Spectroscopic invention fostered the comparison

---

[110]The American representative was August Pfund of Johns Hopkins. The eighty-five wavelengths agreed upon were considered "secondary standards." The system of wavelengths was anchored by a measurement of one cadmium line (6438.4696 angstroms) by French physicists Fabry and Perot. C.G. Abbott, "Recent Progress in Astrophysics," in *Annual Report of the Board of Regents of the Smithsonian Institution, 1913* (Washington, DC: G.P.O., 1914), 177-178.

[111]"Edward Charles Pickering's Diary of a Trip to Pasadena to Attend Meeting of Solar Union, August 1910," ed. Howard Plotkin, in *Southern California Quarterly*, 60 (1978): 35-36.

[112]F.J.M. Stratton, "International Co-operation in Astronomy: A Chapter of Astronomical History," *Monthly Notices of the Royal Astronomical Society*, 94 (1934): 367.

[113]Kevles, "'Into Hostile Political Camps': The Reorganization of International Science in World War I," *Isis*, 62 (1971): 56. Idem, "George Ellery Hale, the First World War, and the Advancement of Science in America," *Isis*, 59 (1968): 435.

[114]W.W. Campbell and Joel Stebbins, "Report on the Organization of the International Astronomical Union," *PNAS*, 6 (1920): 352

of laboratory and solar spectra; that very comparison was guided by the hope of conserving the principle of Kirchhoff and Bunsen; and an awareness of physics as practiced by an international array of spectroscopists provided the groundwork from which the School of Light arose. Physics as practiced at Hopkins was not necessarily simply a way of amassing wavelengths-cum-facts. It was also a process of discovery, and the manner in which research was conducted carried a distinctive imprint. Spectroscopy in the School of Light carried with it an expression of values.

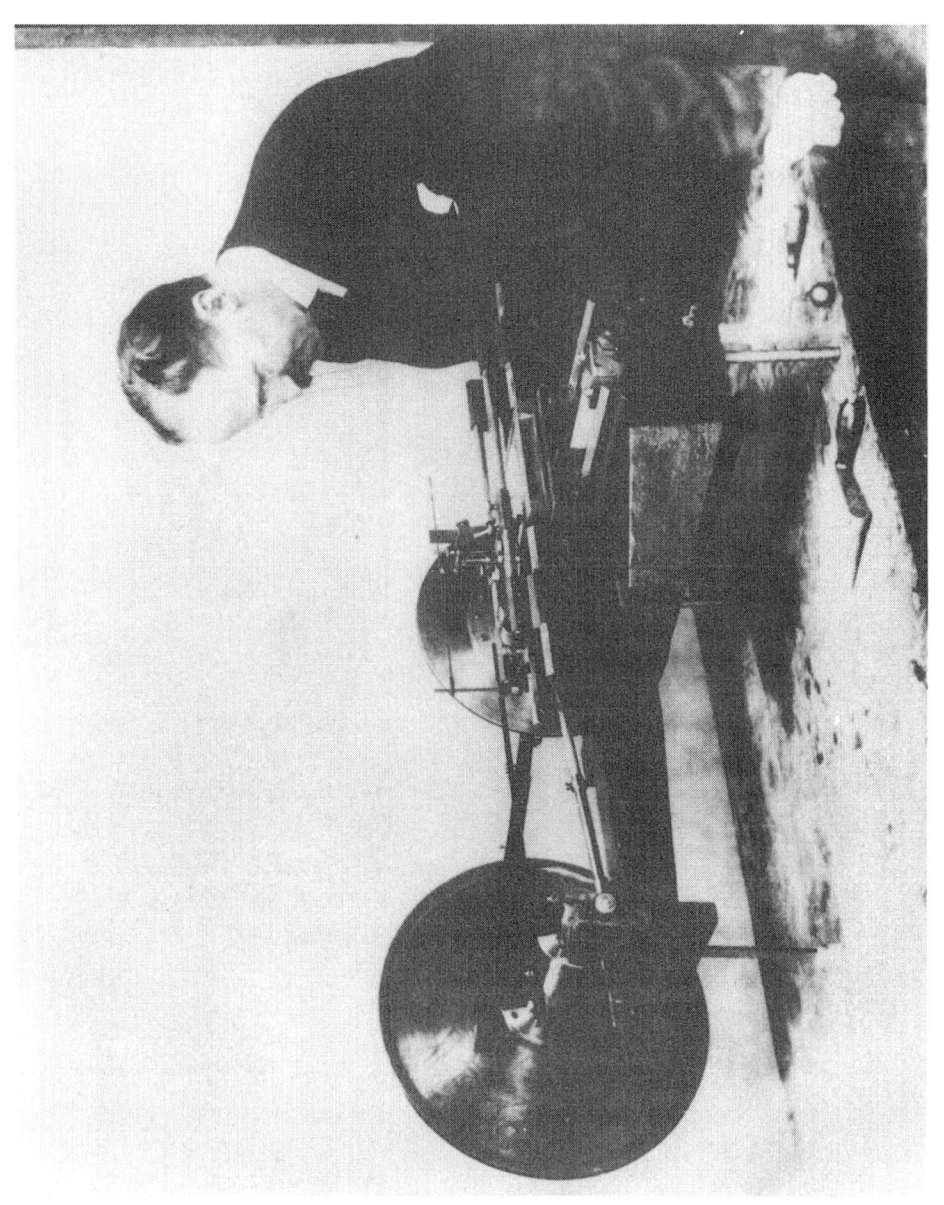

Plate 2. Henry Rowland. Courtesy of the Johns Hopkins University Special Collections.

The Proliferation of Spectral Lines

Despite their conservatism, American experimentalists of the extended Hopkins school actually created a need for more advanced physical concepts; ideas about the atom as they arose abroad were not devoid of American ingredients. The power of experiment inhered not only in the confirmation or refutation of existing concepts, but also in the requirement of conceptual innovation. Like the sand in the shell of an oyster that leads the organism to manufacture a pearl, the plenitude of spectral lines represented objects to be gathered into some smooth conceptual whole. The discipline of spectroscopy produced numbers, denominated usually in angstroms, containing typically four to seven significant digits. These were the phenomena to which putative noumena could be attached. They were the multifarious constituents of light, analyzed physically and instrumentally before they were analyzed mathematically and theoretically.

Ever-improved spectroscopes drove the process. They magnified spectra just as telescopes magnified astronomical objects. Grating manufacturers pursued high dispersion along with durability in the face of the corrosive gases sometimes used in near-vacuum spectroscopy. Even while atomic models came into being, the intellectual environment for them became more demanding, as the number of phenomena requiring explanation grew. For example, in 1912 Robert Wood warned: "The complexity of the iodine absorption spectrum has been greatly underestimated." His own estimate came from state-of-the-art facilities. In a barn at a summer home in East Hampton, Long Island, Wood constructed a plane-grating spectrograph that he believed to be, with the possible exception of a Michelson instrument in Chicago, "the largest and most powerful in the world."[1]

It had been assumed that the iodine spectrum was well-established, but like a microscope switched to higher magnification, Wood's grating went further, and demonstrated that compilation of the iodine spectrum was not yet finished. He wrote, "I find . . . that with every increase in the resolving power that I bring to bear, more structure comes into view." Analyzing one absorption line shown on a previous map, Wood found seven. He reported, "I have employed the highest resolving power that has ever been brought to bear upon the spectrum, and it is still not completely resolved."[2] Wood sought insights from iodine that would be applicable to all elements; the "chemical ferocity" of sodium vapor had led him to switch to iodine.[3] Whatever the element, Wood pursued a fundamental goal: "the nature of the very complicated electro-magnetic mechanism we call the molecule." (Wood took "molecule" to mean the smallest possible particle of matter.)[4]

The trend of experimental and observational spectroscopy was clear: more and more lines, finer and finer detail in the spectrum of each element. Any theory that explained one line might later have to explain seven. Europeans as well as Americans took part in this multiplication. Gratings from Johns Hopkins continued to drive experiment toward the recognition of greater complexity. Other spectroscopic devices, such as interferometers, also contributed to the proliferation.

Contemporaneously with Wood, Michelson observed the same trend toward proliferation of spectral lines: "Rowland's exquisite maps had shown many of these, which were then thought simple, to be double, triple, or multiple, and there are clear indications that even the simpler lines showed differences in width, in sharpness, and in symmetry."[5] And the process had continued after Rowland. Better instruments brought about the detection of finer detail; laboratory work answered questions (what is the spectrum of hydrogen?), but it also raised questions (what law does that spectrum obey?). Complexity had to be attributed to some property of the emitting matter. As Michelson wrote, although wavelengths were still the principal quantitative data obtained from spectra, other characteristics of spectral lines were noted as well,

---

[1]Wood, "Resonance Spectra of Iodine by Multiplex Excitation," *Philosophical Magazine*, 24 (1912): 673-674, 677.

[2]Ibid., 673-674, 677-678.

[3]Wood, "The Optical Properties of Metallic Vapors," 98; "Resonance Spectra," 675.

[4]Wood, "The Optical Properties of Metallic Vapors," 96.

[5]Michelson, "Recent Progress in Spectroscopic Methods," *Nature*, 88 (1912): 363.

for example, whether they are intense or faint, nebulous or sharp, narrow or broad, symmetrical or unsymmetrical, reversed, &c.— characteristics which we recognize to-day as of the highest importance, as giving indications of the structure and motions of the atoms the vibrations of which produce these radiations.[6]

Michelson was, like Rowland, a physicist representative of his nation and age, and both devoted more attention to the spectral lines than to the atomic models to which they could be attributed. Unlike Rowland and Wood, Michelson made important use of interferometers in studying spectral lines. But, like them again, Michelson achieved "the resolution of many lines supposed single into doublets, quadruplets, &c."[7]

Michelson found that while, with an interferometer, the "resolving power is practically unlimited," in practice the "difficulties in the application of the interferometer method" were formidable enough to require recourse to gratings. In contrast to the interferometer and other new devices, the diffraction grating possessed "so many advantages in simplicity and convenience of manipulation" that it could be used to advantage in most situations. Michelson's respect for the diffraction method was most evident in his attempts to sustain grating production in Chicago. He acknowledged in 1912 that Rowland gratings had "been practically the only ones in service" in the previous 30 years, but continued his efforts to provide an alternate source.[8]

In point of spectroscopic principle, Michelson differed with the late Rowland mainly on the probable outcome of progress in instrumentation. In contrast to Rowland who, Michelson said, believed "that the width of the spectral lines themselves was so great that no further 'resolution' was possible," Michelson maintained that greater resolution would bring more discovery. Atmospheric turbulence limited the performance of astronomical telescopes, but laboratory instruments did not face the same limitation, for

> there is no corresponding limit to the effective power of spectroscopes, and the solution of the corresponding problems of the subatomic structures and motions of this ultramicroscopic universe may be confidently awaited in the near future.[9]

Like Rowland, Michelson believed that spectra displayed a code from nature, and that the task of the experimentalist was to make the code plain. To do so required ingenuity and skill. But with a few exceptions, the experimentalists were not cryptographers who would

---

[6]Ibid.
[7]Ibid., 364.
[8]Ibid., 363, 364.
[9]Ibid., 365.

break the code and read a description of the atom. "Our present duty is to make it possible to receive and record such messages." The spectroscopists also decompressed the message, so that the terms no longer blended together, and remained confident that the theorist-cum-cryptographers would extract from their wavelength data a clearer knowledge of the microphysical world.[10]

From at least the 1860s, spectroscopy provided evidence of some inward complexity of matter, and those who engineered solutions to the problem of obtaining high-dispersion spectra simultaneously created problems for theories of matter. Norman Lockyer's chagrin at this process in the 1880s, when it brought difficulties for his theory of molecules, suggests that there were long-standing disincentives to fanning the spectrum out ever more widely. By 1917 Robert Millikan, then at the University of Chicago, could testify to the problems that spectroscopy forced on physics:

> . . . during all the years in which the dogma of the indestructible and indivisible atom was on the stage, it was the complexity of spectra even of simple gases which kept the physicist in the path of truth, and caused him continually to insist that the atom could not be an ultimate thing, but rather that it must have a structure and a very intricate one at that—as intricate, in Rowland's phrase, as a grand piano.[11]

At Millikan's writing, the mechanisms of the "piano" were still poorly understood, although the Bohr atom had by then offered new insight. His appraisal of the compelling effect of spectroscopy on atomic theory holds significance for his own intellection as well as the state of physics: Millikan began to show a greater interest in laboratory spectroscopy, perhaps in proportion to the salience of new atomic theories, and the reliance of those theories on spectral data.

Although Americans did not lead the way in modeling the "piano," many were interested in the process as it unfolded in Europe. Crew in 1907 discussed the "Saturnian atom" that emerged largely at Cambridge University, and thought that this model, partly a product of William Thomson, offered a good fit with experimental data. However, greater dispersion and resolution of spectra promised still greater complexity for any model: "spectral lines are *not* perfectly sharp, but (within limits not yet resolved by any grating) possess a complicated structure." This outlook, this probability that more difficulties would emerge in the form of lines-cum-phenomena, made Crew hold back from a full endorsement of the Cambridge atom, for a theory not only

---

[10]Ibid., 365.
[11]Millikan, "Radiation and Atomic Structure," 194.

had to describe, but also to predict, and theories of the Saturnian atom "appear to be singularly devoid of that spirit of prophecy which characterizes all sound theory."[12] (Where Crew desired an atomic model with a potential for "prophesy" in experiment, Millikan spoke of "fertility" as a desideratum.)[13] Crew's judgment on the Cambridge atom was pragmatic; although it fit observations well, it apparently could not guide experiment to new discoveries. However, from the attention Crew devoted to the subject, it was evident that *some* form of atomic theory that would hold up against the multiplication of spectral phenomena was in demand. And in atomic theory, as in cultural affairs, Americans looked to Europe for sources of authority, although atomic models arose in the U.S. as well.[14]

For physicists of the extended Hopkins school of spectroscopy, spectral lines were fundamental all along; the lines multiplied under high dispersion. In 1913, the only two known spectral series of hydrogen were the Balmer series and the infrared series observed by Paschen with his Hopkins-made instrument (n=2 and n=3, respectively).[15] In ensuing years, other series were adduced. First came Lyman's detection of the first members of the n=1 ultraviolet series, using a Rowland concave grating. Lyman thus increased the number of known series of hydrogen lines to three. Then in 1921 came the dramatic extension of the known Balmer series to 20 terms, by Wood. In 1922, Hopkins physicist F.S. Brackett added a fourth series (n=4), by observing a long hydrogen tube end-on, as in Wood's research. He detected the first two lines in a series in the infrared. Brackett also added three terms to the two terms Paschen had detected in the n=3 series.[16]

Another Hopkins physicist, A.H. Pfund, then took up the extension of observation. Using a rock-salt spectrometer, Pfund, too, studied hydrogen in the infrared; his intent was "to search for a spectral series of hydrogen beyond that discovered by Brackett." Pfund used a thermopile with "a general sensitivity several times as great as that employed by Brackett." The intensity of light was measured by the deflection of a galvanometer. But observation was difficult. Pfund tentatively found one line far out at 7.4 microns, where theory predicted 7.46 microns. Pfund thought the agreement adequate: "no great

---

[12]Crew, "Fact and Theory in Spectroscopy," 4, 7, 11.

[13]Millikan, "Radiation and Atomic Structure," 204-205.

[14]Gerald Holton, "On the Hesitant Rise of Quantum Physics Research in the United States," in *The Michelson Era in American Science, 1870-1930*, ed. Stanley Goldberg and Roger H. Stuewer (New York: American Institute of Physics, 1988), 189.

[15]Bohr, "On the Constitution of Atoms and Molecules," 9.

[16]F.S. Brackett, "A New Series of Spectrum Lines," *Nature*, 109 (1922): 209. Brackett also carried out molecular infrared spectroscopy. Yakov M. Rabkin, "Technological Innovation in Science: The Adoption of Infrared Spectroscopy by Chemists," *Isis*, 78 (1987): 38.

accuracy being claimed for wave-length measurements—it seems probable that the line found is the first member of the new series."[17]

In addition to the establishment of the basic series lines of hydrogen, there was the ramification of those lines at high dispersion. First came the "fine structure" of the hydrogen spectrum, the observation that lines in the Balmer series, when examined at sufficiently high dispersion, could be resolved into two lines, a "doublet." Then came complex structure, a more generic heading, which subsumed the resolution of lines in many spectra into multiple components. Beyond doublets were triplets, quartets, quintets. These groupings included not only sets of lines resolved out of single lines, but also sets of lines that appeared similar in character, often with rules or patterns governing the intervals between them. When in 1922 Miguel A. Catalan in England detected groups of nine and more with "very exact numerical separations," in the spectrum of manganese, he introduced the term "multiplets."[18] This paper "opened the flood gates of complex spectral analysis, and a veritable deluge of publications continued for some years until the main features of arc and spark spectra of most elements (excepting rare earths) were disclosed and interpreted."[19] Still higher spectrographic dispersions led to the recognition that the individual components of multiplets could themselves be resolved into sub-components, under the name of hyperfine structure.[20] And in addition to these complexities in spectra of unperturbed atoms, there were complexities caused by placing atoms in a magnetic field—the Zeeman effect—and in an electrostatic field—the Stark effect. The net result was a proliferation of observations, to which order gradually came.

Hopkins instruments played a driving role in this process. Both the Zeeman effect[21] and the Stark effect[22] were detected with the help

---

[17]A.H. Pfund, "The Emission of Nitrogen and Hydrogen in the Infrared," *Journal of the Optical Society of America*, 9 (1924): 193-196.

[18]Miguel A. Catalan, "Series and Other Regularities in the Spectrum of Manganese," *Philosophical Transactions of the Royal Society of London, Series A*, 223 (1923): 147, 168. Catalan worked in the laboratory of Alfred Fowler at Imperial College, South Kensington. Fowler's mentor had been Norman Lockyer.

[19]Meggers, "Spectroscopy, Past, Present, and Future," *Journal of the Optical Society of America*, 36 (1946): 432.

[20]Ibid., 436. Gerhard Herzberg, *Atomic Spectra and Atomic Structure*, trans. J.W.T. Spinks (New York: Dover Publications, 1945; first published in English 1937), 182. Some of the classification of hyperfine structure was carried out by Arthur Ruark (Johns Hopkins Ph.D., 1924). Ruark, Manuscript biography 10 (1963), Niels Bohr Library, American Institute of Physics, College Park, MD, p. D-2.

[21]P. Zeeman, "On the Influence of Magnetism on the Nature of the Light Emitted by a Substance," *Philosophical Magazine*, 43 (1897): 227.

[22]J. Stark and H. Kirschbaum, "4. Observations of the Effect of an Electrical Field on Spectral Lines. IV. Types of Lines, Spreading," *Annalen der Physik*, 43 (1914): 1023.

of Rowland's instrument. Catalan, too, used Rowland's instrument, in addition to a glass prism and a quartz prism.[23] A signal usage of the concave grating occurred in the laboratory of Friedrich Paschen, in the study of the simplest of complex structures, the doublets in the hydrogen Balmer series. The instrument had already served Paschen well. In the 1890s Paschen and Carl Runge used a six-inch Rowland grating to study the spectrum of terrestrial helium.[24] This work of Runge and Paschen apparently made the latter's name: "Overnight Paschen acquired an international reputation."[25] To isolate the new gas, Paschen and Runge had to drive off hydrogen, "the most persistent impurity." To obtain wavelengths, they interpolated from standard iron and sodium wavelengths established by Rowland. In extending the study into the infrared, Paschen standardized a prism apparatus by means of a Rowland grating originally manufactured for Samuel Langley, and loaned to Paschen by James Keeler.[26] The fame was Paschen's, but the hand of the Hopkins school was involved.

In 1899 Paschen requested that a new grating be made, after he returned one to Langley. The grating was to be relatively coarse-ruled, probably for work in the infrared. Because the task would involve readjusting the ruling engine, Paschen hesitated to ask.[27] But the order was filled in 1900, and Paschen thanked Rowland fervently:

> As far as I can judge from looking at the spectra of a sodium flame, the grating is very beautiful in definition and in intensity of the spectra. I thank you very much for your great kindness and I only regret that you had so much trouble in making the changements [sic] in your dividing engine. The grating will be of the utmost value to me for determining greater wavelengths. I am delighted that I can take up this work now with better chances than before.[28]

This instrument likely remained in Paschen's laboratory for some time, but over the years he probably used others from the same source. A Rowland plane grating figured in Paschen's discovery of a third spectral series of hydrogen, in the infrared, in 1908.[29] Paschen and Ernst Back also used a Rowland grating in 1912, in their studies of

---

[23]Catalan, "Series and Other Regularities," 128-129.
[24]C. Runge and F. Paschen, "On the Spectrum of Cleveite Gas," *ApJ*, 3 (1896): 6 note. Cleveite was a mineral which yielded helium.
[25]Paul Forman, "Paschen, Louis Carl Friedrich," *DSB*, Vol. X (1974), 346.
[26]Runge and Paschen, "On the Spectrum of Cleveite Gas," 5, 23.
[27]Paschen to Rowland, 31 March 1899, 5 October 1899, HAR.
[28]Paschen to Rowland, 9 February 1900, HAR.
[29]Paschen, "Zur Kenntnis ultraroter Linienspektra," *Annalen der Physik*, 27 (1908): 543-544.

the "normal" and "anomalous" Zeeman splitting of lines in magnetic fields of differing strength.[30]

Thus by the time Paschen turned to fine structure, he knew the instrument well. Paschen did not discover fine structure; the observation that individual lines of the hydrogen spectrum can be resolved into pairs of lines predated 1900.[31] But when Arnold Sommerfeld developed a model that could explain fine structure in 1915-1916, Paschen provided experimental confirmation. Paschen worked with helium, for which the separation of fine-structure components was greater than hydrogen. Part of his claim to superiority in measurement derived from his use of a Rowland instrument.[32] Presumably that instrument was a scarce resource in Germany because of hostilities, and because of limited production in Baltimore.

Employment of this instrument, without careful experimental technique, still could not guarantee the last word. To complete a usable exposure of a few hours, Paschen had to maintain great mechanical rigidity in his overall apparatus, as well as constant temperature. While war raged elsewhere, Paschen engineered a stillness in his laboratory. The results, Paschen felt, were definitive and unequivocal.[33]Unlike the work of Lyman and the formula-directed research of Wood, Paschen's studies were very much theory-directed, and he was quite interested from the start in assimilating his results to Sommerfeld's model.[34] Later measurements by other spectroscopists showed disagreement,[35] but in the long run Sommerfeld's energy formula for the hydrogen atom held up. As Helge Kragh writes: "By some sort of historical magic, Sommerfeld managed in 1916 to get the correct formula from what turned out to be an utterly inadequate model."[36] Atomic models continued to evolve in subsequent years, and in this process Paschen's work provided challenges as well as support. Closure of fine-structure investigations occurred in 1926 or soon thereafter, according to Kragh.[37]

---

[30]Paschen and E. Back, "Normale und Anormale Zeemaneffekte," *Annalen der Physik*, 39 (1912): 900, 906. Later investigators who extended this research reported using an echelon grating because they did not have a Rowland grating at their disposal. K. Forsterling and G. Hansen, "Zeemaneffekt der roten und blauen Wasserstofflinie," *Zeitschrift fur Physik*, 18 (1923): 27.

[31]At about the same time as their "ether drift" experiment, Michelson and Edward Morley found, with an interferometer, that "the red hydrogen-line must be a double line." Michelson and Edward W. Morley, "On a Method of Making the Wave-length of Sodium Light the Actual and Practical Standard of Length," *Philosophical Magazine*, 24 (1887): 466.

[32]Paschen, "Bohrs Heliumlinien," *Annalen der Physik*, 50 (1916): 903, 906-907.

[33]Ibid., 903.

[34]Helge Kragh, "The Fine Structure of Hydrogen and the Gross Structure of the Physics Community, 1916-26," *Historical Studies in the Physical Sciences*, 15 (1985): 73.

[35]Ibid., 69.

[36]Ibid., 84.

[37]Ibid., 115, 122.

Significantly, the measurements that in Kragh's view resolved the physical picture were finished at Caltech in Pasadena, with its ruling operation. William Houston started his experiments at Ohio State University; Norton Kent and colleagues started theirs at Boston University.[38] But both lines of research ended in Pasadena. Per Kragh, "With the experiments of Houston and Kent consensus was finally achieved about the fine structure of the hydrogen spectrum."[39]

Beyond fine structure, hyperfine structure challenged theorists to extend their models deeper into the atom, inside electron orbitals, to the atomic core. Nuclear mechanical and magnetic moments served to explain the very minute spectral line separations. Observations of hyperfine structure required "the sharpest lines and greatest spectroscopic resolving power." Examples were known by 1925; by about 1930 both interferometric devices and large concave gratings were brought to bear on measuring line separations.[40] One major study involved a large grating spectrograph at Mount Wilson and was written up at the Physikalisch-Technische Reichsanstalt in Berlin.[41]

At the level of experimental practice, the flow of laboratory expertise was not unidirectional. When Paschen was pursuing fine-structure studies in Germany, Meggers in the U.S. asked Lyman how he could best mount his concave gratings, and Lyman answered that he used a mounting developed in Germany by Paschen. ("Matters are arranged so that the whole apparatus is in a light-tight room all the manipulation in front of the slit being carried on in a separate chamber.")[42] The instrument had to be protected from heat, vibration, and extraneous light. Paschen's arrangements had come a long way since 1895, when Henry Crew discovered him using a concave grating with no proper mounting at all.[43] Still later, in 1925, Meggers wanted to verify studies by Paschen and wrote to him, "Unless you can suggest any way to improve on your own arrangement and procedure we are going to copy it as closely as we can."[44]

Theorists lived in a different domain, where knowledge of the ways of their colleagues in experiment was often tacit and was not always within the purview of their printed discussions. For Sommerfeld,

---

[38]William V. Houston, "The Fine Structure and the Wave-Lengths of the Balmer Lines," *ApJ*, 64 (1926): 92. Norton A. Kent, Lucien B. Taylor, and Hazel Pearson, "Doublets Separation and Fine Structure of the Balmer Lines of Hydrogen," *The Physical Review*, 30 (1927): 283.

[39]Kragh, "The Fine Structure," 122.

[40]Meggers, "Spectroscopy, Past, Present, and Future," 436-437. Herzberg, *Atomic Spectra and Atomic Structure*, 182.

[41]H.E. White, "Hyperfine Structure in Singly Ionized Praseodymium," *The Physical Review*, 34 (1929): 1397. At very high dispersion, some faint ghosts re-emerged. Ibid., 1398.

[42]Lyman to Meggers, 25 July 1916, TL.

[43]Crew, "Diary," 12 July 1895, quoted in Forman, "Paschen, Louis Carl Friedrich," 346.

[44]Meggers to Paschen, 28 August 1925, WFM.

the most important natural objects were the unobservable constituents of the atom, not the manipulable instruments of the spectroscopic laboratory. Paschen's measurements showed, Sommerfeld asserted, that non-relativistic theories of fine structure were necessarily wrong.[45] In exposition, Sommerfeld nearly removed the experimenter and his equipment from the evolving picture:

> The most beautiful and instructive manifestation of the various elliptic orbits that belong to the same Balmer line is . . . given by Nature herself without our agency in the fine structure of space-time conditions as reflected in the fine structure of spectral lines.[46]

Sommerfeld elided the role of the experimenter and the theorist in obtaining confirmation from nature. He knew of Rowland's instrument, but minimized the human agency residing in it. He simply accepted that Paschen's apparatus offered a clear glimpse of atomic emissions; he was not troubled by, for example, the possibility of ghosts that haunted American physicists. He accepted what Lyman could not, namely that Rowland gratings were "of perfect construction."[47] In the very process of accepting Rowland gratings as "perfect," Sommerfeld overlooked human agency. With the instrument in his estimation perfected, he saw not flaws but nature.

Sommerfeld also saw nature imitating art. Rowland's instrument was not usable for the very short wavelengths of x-ray spectroscopy, but Max von Laue realized that the atomic lattices of crystals possessed the proper dimensions to diffract x-rays. Use of crystals as x ray diffraction instruments gave information on atomic lattices as well as x-rays. Sommerfeld thought it a masterstroke:

> The brilliance of Laue's idea consisted in his recognizing that the space-structure of crystals is just as happily adapted to the wavelength of Rontgen [sic] radiation, as the structure of a Rowland grating was adapted to the wave-length of ordinary light, that is, that we can take directly out of the hands of Nature the diffraction apparatus necessary for Rontgen [sic] rays.[48]

Here, products of nature were just as good as products of the Baltimore engine. For the most part, mentions of Rowland's device were few and far between in Sommerfeld's work. This contributed to the seeming naturalness of his theorizing. But Sommerfeld the theoretician maintained an active dialogue with Paschen the experimentalist.

---

[45]Arnold Sommerfeld, *Atomic Structure and Spectral Lines*, trans. Henry L. Brose (London: Methuen & Co., 1923), 531.

[46]Ibid., 237.

[47]Ibid., 113.

[48]Ibid., 117.

The initial fine structure of the hydrogen spectrum was only the beginning of the discovery of fineness in the structure of many spectral lines in many elements. The hydrogen doublets were not alone. Meggers worked with a variety of materials, and by 1924 detected not just doublets, but triplets, quintets, and septets, if not groups of nine like Catalan's.[49] Complex structure came to mean any inward complexity of the individual lines in spectral series, and initially, in the 1910s, "no one really knew to what physical phenomenon to attribute the presence of complex structure."[50] In Paul Forman's view, Sommerfeld became discouraged, in 1919, with his *a priori* approach to explaining spectral lines, trying to deduce them from given atomic models, and turned instead to an approach that was, rather, *a posteriori*, deducing atomic features, especially the energy levels of orbiting electrons, from spectral lines. Sommerfeld and his students employed this methodology with both optical and x-ray spectra in the years 1919-1921.[51]

Although exposition usually places the atom first and the light from the atom second, the historical sequence of discovery was at times the reverse. "It began with the observed spectral lines, and worked back to the energy levels. These levels were then characterized by quantum numbers and selection rules—invented *ad hoc* if necessary."[52] This was the method in theoretical practice, not necessarily in theoretical exposition. Sommerfeld's importance to physics derived partly from his book, *Atomic Structure and Spectral Lines*. Meggers recalled, "Spectroscopists were amazed that our meager knowledge of atomic structure and the origin of spectra could be expanded into such a big book."[53]

After World War One, Meggers took a leading role in reestablishing communications between American and German physicists, especially with Sommerfeld. Meggers planned a new map of the infrared solar spectrum and better chemical identification of solar lines.[54] He also felt that the most important task facing spectroscopy was to describe the different types of spectra (arc, spark, and discharge tube) of all the chemical elements as completely as possible, for specific conditions.[55] Later, Meggers demonstrated the immensity of the task facing spectroscopy by showing that, if each element possesses one spectrum for each electron, then 92 elements will have 4,278

[49]Meggers to Otto Laporte, 23 January 1924, WFM.
[50]Forman, "Alfred Lande and the Anomalous Zeeman Effect, 1919-1921," *Historical Studies in the Physical Sciences*, 2 (1970): 185.
[51]Ibid., 186.
[52]Ibid.
[53]Meggers, "Spectroscopy, Past, Present, and Future," 431.
[54]Paul W. Merrill to Meggers, 23 September 1919, WFM.
[55]Meggers to Charles E. St. John, 19 January 1920, WFM.

spectra in total,[56] each with its various series and complex makeup, requiring separate analysis.

With the prospect of so much complexity, Meggers had reason to desire organizing principles, which Sommerfeld could likely provide. For his part, Meggers made sure that Sommerfeld received an American optical journal, and that Sommerfeld knew that his book *Atomic Structure* provided "much inspiration for experimental work."[57] When a new edition was pending in 1922, Meggers informed the author: "We have a ravenous hunger for your work."[58] In the spring of 1922, Meggers made inquiries on Sommerfeld's behalf to several universities, asking if they wanted Sommerfeld to lecture. Lyman at Harvard agreed to four lectures.[59]

Meggers the civil servant took an activist part in the resumption of the diplomacy of science in the aftermath of World War I. In 1921, he made a twelve-week tour of Europe, meeting spectroscopists and other physicists in various nations, in a trip reminiscent of Hale's earlier travels. In Germany, Meggers found many physicists unable to afford foreign scientific journals and "diffident in initiating correspondence, fearing unpleasant rebuffs, and perhaps a reminder that they are responsible for the world war."[60] After returning to the U.S., Meggers redressed the former problem by arranging free subscriptions to the *Journal of the Optical Society of America* for German research centers.[61]

Although Meggers's domain was very much the laboratory, he remained aware that physics and astronomy had much to share, and maintained contact with spectroscopists at Mount Wilson Observatory, and with astronomer Henry Norris Russell at Princeton. Only astronomy had emerged unblemished from the war:

> It is my opinion that Astronomy is our nearest approach to sublime pure science. It had little if any direct responsibility for the recent terrible and inhuman war to which Physics and Chemistry prostituted themselves.[62]

While Germans were still being ostracized in many cases, in 1922 Meggers favored the admission of Germany and Austria to the International Astronomical Union.[63]

---

[56]Meggers, "Spectroscopy, Past, Present, and Future," 432. At Meggers's writing in 1946, only 101 spectra were "in a state of practically complete description and interpretation." Ibid., 436.
[57]Meggers to Sommerfeld, 9 December 1921, WFM.
[58]Meggers to Sommerfeld, 24 February 1922, WFM.
[59]Lyman to Meggers, 26 May 1922, TL.
[60]Meggers to Alfred Fowler, 6 November 1921, WFM.
[61]Meggers to H.M. Konen, 9 November 1921, WFM.
[62]Meggers to Joel Stebbins, 12 November 1921, WFM.
[63]Meggers to J.M. Eder, 28 March 1922, WFM.

German syntheses of spectroscopic work remained in demand. Meggers tried to arrange National Research Council funding for a new volume of Heinrich Kayser's encyclopedias of spectroscopy, which provided a critical summary of research. When the $1,000 Kayser asked was not forthcoming, Lyman offered to furnish the amount personally, provided that he would later be repaid.[64] From the Old World, Meggers also hoped to acquire pure materials, and a concern with purity here echoed Lyman's desideratum in earlier studies of hydrogen. Before his 1921 trip, Meggers told a Bureau associate, "Perhaps I can find some *pure* Helium, Argon, Krypton and Xenon in Europe." In 1922, the Bureau obtained three bulbs of pure helium gas from Kamerlingh Onnes of Leiden, which was prized "very highly." From physicists in Denmark it acquired in 1925 a sample of compounds of very pure hafnium.[65]

In the early 1920s, according to Meggers, "the spectroscopic work of the Bureau of Standards changed its complexion somewhat, formerly we were content to describe spectra, but more recently we have become fascinated with the search for spectral regularities."[66] The search for regularities was not necessarily theory, but it did represent an attempt to assimilate novelty and variety, in the form of spectral lines, to schemas familiar from spectra already classified. The regularities Meggers sought were primarily groups of spectral lines with similar character and appearance. Research at the Bureau approached theory, but still gave primacy to that which, unlike the atom, could be photographed and measured directly. The approach was phenomenological. Meggers wrote to a fellow spectroscopist at Mount Wilson that he and Carl Kiess, a colleague at the Bureau, were "still struggling with the structures of complex spectra." The answer was more measurements: "So long as we stay by this task there is little danger that we will have no more to do."[67] What accumulated was not necessarily anomaly, but novelty.

Although Meggers was, like Hale, a research scientist, he did not have the support of private philanthropy, and had to perform practical work to protect the existence of his section of the Bureau. This work benefitted from the know-how Meggers gained in the pursuit of pure research, which was sometimes "said to be a waste of funds." Meggers's section of the Bureau justified its work publicly by ascertaining the composition of materials. He wrote to a colleague, "The

---

[64]Meggers to Keivin Burns, 3 May 1921; Lyman to Meggers, 23 June 1921, WFM.
[65]Meggers to C.C. Kiess, 2 July 1921; Meggers to H. Kamerlingh Onnes, 28 March 1922; Niels Bohr to Meggers, 11 September 1925, 23 October 1925, WFM.
[66]Meggers to Alfred Fowler, 28 August 1925, WFM.
[67]Meggers to Merrill, 25 November 1925, Paul Willard Merrill Papers (PWM), Huntington Library, San Marino, CA.

only effective excuse we have for doing any work in spectroscopy is the fact that each year we make several hundred spectro-chemical analyses of complex or defective materials representing industrial problems which the ordinary chemical methods are not able to solve."[68] The section carried out tasks for other government departments. It assessed the purity of gold for the U.S. Mint, and developed photographic techniques for the War Department.[69] Despite these services, government support for the section remained under pressure in the 1920s.[70] Meggers said his section had a "small staff of 3 people, but we all work day and night and holidays."[71]

Within the discipline of spectroscopy, Paschen confirmed Meggers's status as an outstanding experimentalist. Although Paschen had discovered the second known spectral series of hydrogen, and made important early studies of fine structure in spectral lines, he felt in 1924 that he could not offer some American physicists anything new. Still, he planned to lecture in the U.S., and of all the laboratories he hoped to visit, he wanted above all to see the spectroscopic laboratories of the Bureau of Standards. In Paschen's view, Meggers developed experimental physics there as nowhere else.[72]

Among the first triumphs in the search for patterns in spectra was a relation between elemental spectra and the position of a given element in the periodic table. Radiation from ionized atoms (spark spectrum) of a given element was found to resemble radiation from neutral atoms (arc spectrum) of the preceding element in the periodic table. Also, the multiplicities, or number per group, of spectral lines alternated, for successive elements, between even and odd numbers. These rules existed before Meggers, but his observations bore them out. He found that in alkalis, "energized states are double," while in alkaline earths the states could be "either single or three fold."

Multiplicities correlated with the number of electrons in the outermost orbit of an atom. Meggers felt that most observations were "in beautiful accord with the quantum theory of spectral line emission." Between 1922 and 1925, he reported, more than 5,000 lines had been classified by just seven workers, including some at the Bureau of Standards. However, there were very complex spectra that "resisted practically all attempts to arrange the lines in regular sequences."[73]

---

[68]Meggers to Henry Norris Russell, 14 June 1924, WFM.
[69]Meggers to Director of the Mint, Washington, DC, 4 March 1920, 23 March 1920; Meggers to Russell, 14 January 1924, WFM.
[70]Meggers to Merrill, 8 November 1926, PWM.
[71]Meggers to Russell, 25 July 1923, WFM.
[72]Paschen to Meggers, 14 June 1924, WFM.
[73]Meggers, "Periodic Structural Regularities," 44-45.

Still, at the outset of 1925 there were spectroscopic phenomena for which "no theoretical laws or suggestions" existed. Here, "theory" generally meant atomic models. Meggers felt that there were still relations between the azimuthal quantum numbers and energy diagrams that remained to be articulated. He also looked for explanations of two of the most significant spectral lines in an elemental spectrum, the *raie ultime* and the resonance line. The *raies ultimes* were "the most sensitive lines for spectrographically detecting the chemical element." When chemicals were analyzed by spectroscopy, the *raie ultime* of a given element was the most certain to appear. Meggers stated, in modern language, that the line arose from a transition to the lowest level from the first higher "non-metastable" level. (He actually wrote of "combination" rather than "transition," and of "term" rather than "level.")[74]

Meggers used the term "resonance" differently from Robert Wood: a resonance line was the line of greatest wavelength in a spectrum, possessing the least energy. Meggers stated that the resonance line resulted from a transition (combination) between the level (term) of the highest multiplicity (possessing the greatest number of sub-levels) in a spectrum and the lowest level. For elements in which the normal state also possessed the highest multiplicity of any state, the *raie ultime* and the resonance line were the same. Meggers did not have the benefit of some quantum-theoretical principles even then being developed, so there were atomic properties, accepted soon thereafter, that he did not invoke. In this characterization of key spectral lines, Meggers merely wished "to call attention to a few empirical rules which perhaps will give some guidance for later theoretical investigations." Atomic theory was developing rapidly, however, and by the time the article went to press, Meggers believed that contemporaneous work by Wolfgang Pauli, Werner Heisenberg, and Friedrich Hund provided theoretical explanation for his observational rules regarding *raies ultimes* and resonance lines.[75] The theory of spectra was still short of completeness, however, as evidenced by Meggers's expression of confidence to a colleague: "Some of these days the theory will be the best criterion for the analysis of an unclassified spectrum."[76]

Although complex spectral structure created challenges for theory, spectroscopy provided instances of support, as well. Although explanations of complex spectra required refinements of Bohr's theory, that theory received important vindication in observations of

---

[74]Otto Laporte and Meggers, "Some Rules of Spectral Structure," *Journal of the Optical Society of America and Review of Scientific Instruments*, 11 (1925): 459-460.

[75]Laporte and Meggers, "Some Rules," 462-463, 463 note.

[76]Meggers to Henry Norris Russell, 4 November 1925, WFM.

simpler spectra. By the mid-1920s, the extended Hopkins school, including Paschen with his Rowland grating, had provided important confirmation for Bohr. Spectroscopic observations were essential to the viability of his theory: "When Niels Bohr published his atomic theory in 1913, most physicists considered its treatment of the hydrogen spectrum its most impressive and satisfactory part."[77]

While European theorists furnished models that were in demand by American spectroscopists, the latter group also possessed something desired by theorists: their assent. It was not automatic. Americans were not always uncritically accepting toward quantum theory. Reservations were sometimes expressed. A paradoxical appraisal came from spectroscopist Paul Merrill at Mount Wilson Observatory. H.A. Lorentz and Paul Epstein spoke in California on quantum theory, and Merrill gained the impression "that the foundations are unsatisfactory, inconsistent and practically incredible. Nevertheless great truths are contained in the theory and it is very useful."[78]

The distinction between theory-oriented Europeans and experiment-oriented Americans can be overdrawn: even in Europe, at the turn of the century, theoretical physicists were a rare breed.[79] As their number increased, they needed phenomena that needed explaining, and more importantly, they needed a place in the physics community. The historical development of physics in the U.S. meant that experiment-oriented physical scientists such as Lyman, Millikan, Hale, and others were, in America, senior. Schweber writes, "In the case of theoretical physicists, experimentalists made the market."[80] (Even into the 1930s, Lyman traveled in England, apparently to collect opinions about physicists who held promise. Thus Hans Bethe was *"excellent but going to Cornell for a year."* George Gamow was "very promising—not a mathematician." As for Erwin Schrödinger: "shot his bolt." And Max Born was "too old. Health?")[81]

In the 1920s, experimentalists knew wavelengths of spectral lines, and intensities. They needed physicists who could make sense of their tables teeming with numbers. According to Schweber, "Professional theorists multiplied because they were useful to experimentalists." Where U.S.-trained theorists existed, especially at Harvard, the partnership was close: "theoreticians not only shared a department with experimentalists, but were also trained in large part by them."[82]

---

[77]Kragh, "The Fine Structure," 68.

[78]Merrill to Meggers, 2 March 1922, WFM.

[79]S.S. Schweber, "The Empiricist Temper Regnant: Theoretical Physics in the United States, 1920-1950," *Historical Studies in the Physical and Biological Sciences*, 17 (1986): 69.

[80]Ibid., 72.

[81]Lyman, "General Notes" (England, 1933, 1934), 5 October 1934, 8 October 1934, TL.

[82]Schweber, "The Empiricist Temper Regnant," 57, 74.

Robert Wood continued his work on the iodine spectrum at Hopkins, but now Edwin Kemble at Harvard held out the hope that molecular quantum theory could help interpret the results. Wood's observations "should provide data from which we can get more accurate values of the moment of inertia and other molecular constants than before."[83] Wood also asked Kemble to suggest new ideas for spectroscopic studies that might be worthwhile.[84] Kemble had taken an interest in analyzing molecular spectra after learning of observations that seemed to call for quantum principles. The presence of lines, rather than continuous bands, in some molecular spectra suggested that classical theories could not give an adequate explanation, and discontinuities had to be introduced.[85]

Kemble's acceptance of quantum principles gave Wood a domestic source for theoretical explanation of spectra. Wood sent Kemble spectrograms, and received insight in return—Kemble employed molecular rotational and vibrational quantum numbers to explain observed lines.[86] This collaboration resembled the Meggers-Sommerfeld relationship, although it was less far-reaching.

Other American partnerships between experimentalists and theoreticians existed. Princeton astronomer Henry Norris Russell fitted the latter category in some instances, and for observational data he drew on Meggers at the National Bureau of Standards, spectroscopists at Mount Wilson, and Hopkins-trained spectroscopist Frederick Saunders, at Harvard.[87] Russell shared Meggers's buoyant optimism about the classification of spectra. Although quantum physics is usually said to have been in crisis prior to 1925, Russell wrote to Meggers in 1923 that he thought "all the more important spectra ought to be cleaned up in a short time." Russell and Meggers agreed that stellar and terrestrial spectroscopy were complementary to each other, and Russell defended Meggers's work to the director of the Bureau. After Russell visited the spectroscopic laboratory, Meggers reported a favorable impression: "We had never thought of you as an intensive spectroscopist and were surprised at the grasp you have on the subject and the marvelous facility you have for finding the laws of spectra." It was clear that the astronomer and the physicist had much to say to each other. The meetings continued, and were an encouragement to

---

[83]Edwin Kemble to Wood, 17 December 1925, Reel 52, AHQP.

[84]Wood to Kemble, 25 December 1925, Reel 52, AHQP.

[85]Holton, "Hesitant Rise," 185-186.

[86]Wood to Kemble, 24 December 1925; Kemble to Wood, 7 January 1926, Kemble to Wood, 14 February 1929, 11 March 1929, Reel 52, AHQP.

[87]Ralph Kenat and David H. DeVorkin, "Quantum Physics and the Stars (III): Henry Norris Russell and the Search for a Rational Theory of Stellar Spectra," *Journal for the History of Astronomy*, 21 (1990): 165.

Meggers's small team. The latter wrote, "We always look forward to these spectroscopic 'pow-wows' because of the suggestions and inspiration which we receive from them." For his part, Russell obtained, from physics, a basis for the interpretation of stellar spectra.[88]

Frederick Saunders, who had studied under Rowland at Johns Hopkins, began collaborating with Russell in the early 1920s, providing first-hand knowledge of laboratory spectra. In correspondence with Saunders, Russell elaborated ideas that led to their joint conceptual and theoretical innovation—the conclusion that two electrons sometimes change state jointly, while absorbing or emitting a single quantum of light. This innovation, soon known as "Russell-Saunders coupling," became their major contribution to quantum physics. It explained the appearance of energies that could not be attributed to the transitions of a single electron.[89]

In their publication on two-electron coupling, Russell and Saunders also attempted a less-remembered task: they sought to bring uniformity to the bewildering variety of symbols used to denote the multitude of spectral lines and energy levels then known. It was an enterprise reminiscent of the earlier attempt of the International Astronomical Union to bring order to the nomenclature for spectral types of stars. Russell and Saunders found "chaos" existing in spectroscopic notation in 1925, with Sommerfeld using one system, Rydberg and Fowler using another, and Bohr using still another. Even Paschen and Sommerfeld differed on some symbols.[90]

In response, Russell and Saunders consulted with colleagues and devised a system that might work until a better arrangement could be established more formally. A new notation system first specified the "serial number" or "total-quantum number," corresponding to the terms in the Ritz formula that gave rise to the Balmer series, Lyman series, and so on. The notation next specified the "series," corresponding to the azimuthal quantum number, using letters inherited from an older spectroscopy: S, P, D, F, G, etc., standing for "sharp," "principal," "diffuse," and "fundamental." The latter designations originated in qualitative observations of spectral lines, predating the arrival of quantum theory. A subscript after the letter assigned each term a number as a component of a multiplet system, and a superscript before the letter specified the number of terms in the multiplet. The subscript also stood for the inner-quantum number. Finally, the

---

[88]Russell to Meggers, 20 July 1923; Meggers to Russell, 25 July 1923; Russell to G.K. Burgess, 21 December 1923; Meggers to Russell, 14 January 1924, 29 April 1926, WFM.

[89]Kenat and DeVorkin, "Quantum Physics and the Stars (III)," 167.

[90]H.N. Russell and F.A. Saunders, "New Regularities in the Spectra of the Alkaline Earths," *ApJ*, 61 (1925): 38, 65.

transition giving rise to a specific wavelength of light was represented by expressing the final and initial terms, respectively. Thus, one line in the calcium spectrum was given by:

$$1\ ^3P_2 - 2\ ^3D_3.$$

(In more modern language, the spectral line was caused by an electron transition from the third in a set of three states in the D orbital of the second shell, to the second in a set of three states in the P orbital of the first shell. Yet this language was not yet that of Russell and Saunders.) The treatment given was phenomenological. Russell and Saunders described spectral lines more than they described atoms.[91]

Russell-Saunders coupling was for both scientists their only major foray into atomic theory. Russell did not remain in terrestrial atomic physics; to some extent he had reached the limit of his ability.[92] After 1925, his interests turned toward the question of elemental abundances in the sun and stars.[93] Saunders, never very theoretically inclined, published little more on spectroscopy, theoretical or experimental. Even while he elaborated on prevailing atomic models with Russell, he expressed dissatisfaction with them. Although aunders indeed saw "a tendency for the electrons to form themselves into pairs," he was skeptical about "the game of atom-building."[94] That the name of Saunders is associated with a theoretical innovation, electron coupling, is rather ironic, given his preference for experiment. He later felt that credit for Russell-Saunders coupling should go to Russell.[95]

Saunders remembered the goal of treating spectra in analogy with music, deducing atomic structure, "just as a blind listener might seek to deduce the fundamental structure of a musical instrument from the tones which it emitted."[96] But the later course of his career casts doubt on whether he thought this goal attained. Sommerfeld, by contrast, thought the music was understood: "What we are nowadays hearing of the language of spectra is a true 'music of the spheres' within the atom, chords of integral relationships, an order and harmony that becomes ever more perfect in spite of the manifold variety."[97] After the joint paper with Russell, however, Saunders published only four more papers on spectroscopy. He turned more to music more strictly

---

[91]Russell and Saunders, "New Regularities," 66.

[92]Kenat and DeVorkin, "Quantum Physics and the Stars (III)," 176.

[93]DeVorkin and Kenat, "Quantum Physics and the Stars (II): Henry Norris Russell and the Abundances of the Elements in the Atmospheres of the Sun and Stars," *Journal for the History of Astronomy*, 14 (1983): 197-207.

[94]Saunders, "Some Aspects of Modern Spectroscopy," 50, 52.

[95]Saunders to Mr. Shipton, 4 February 1963, Harvard University Archives, HUG 4769.200.

[96]Saunders, "Some Aspects," 48.

[97]Sommerfeld, *Atomic Structure and Spectral Lines*, viii.

speaking, specifically to the study of violins, which he pursued for more than two decades. Thus Saunders never gave up on finding music; he simply stopped seeking it among atoms.

In the mid-1920s, Saunders had reservations even about the Bohr model:

> The charm of the Bohr theory is the simplicity which it has introduced into spectroscopy; but so long as the picture it gives us of the atom is complex and full of apparent miracles, the search for simpler conceptions will surely continue.

The model, with refinements by Sommerfeld, was "a beautiful product of the human imagination," but still gave pause. Reconciliation with the static model of chemist G.N. Lewis was needed, as was an explanation for the tendency of electrons to form pairs. Saunders still found it hard to relinquish a medium for electromagnetic waves. And the apparent "knowledge" by the electron of the proper radiation to emit while changing orbits was "less physics than metaphysics."[98]

> Instead of theorists, Saunders favored his own kind, those humble toilers, the experimental spectroscopists, who have all this time been busily grubbing about in the terra firm of observation, and have after all succeeded in raising from their labors the very considerable crop of facts upon which the theory has been nourished.

Saunders singled out one line of research, by Paschen, as perhaps "the most remarkable"—his studies of the neon spectrum showing that probably all spectra, after accounting for complexities, could be understood as series spectra. Although Saunders published little more on spectra, he agreed with Russell that it was "the heroic age of spectroscopy."[99]

Saunders and Meggers "shared the feeling that the substance and theoretical underpinnings of the new mechanics did not have to be mastered in order to use it to interpret laboratory spectra."[100] The case of Saunders does not bear out this feeling; he left the field. Meggers, however, continued to observe spectra for decades, without figuring significantly in quantum theory. His case, especially, shows the subsequent existence of a laboratory-level discipline, remaining close to the phenomena.

As throughout the extended Hopkins school, the stylistic emphasis of those who remained in spectroscopy lay on light as a phenomenon, as a manifestation to be observed and measured. Whether it confirmed theories or created problems for them, it both

---

[98]Saunders, "Some Aspects," 48-50.
[99]Ibid., 48, 50, 53.
[100]Kenat and DeVorkin, "Quantum Physics and the Stars (III)," 177.

represented and was itself nature, and was therefore appropriate for study. Francis Bacon had asserted, "Nature to be commanded must be obeyed."[101] Rowland had stated this approach to nature somewhat differently: "To command her we must know her language."[102] But the language of spectra was enormously complicated, calling for innovation merely to sustain accurate cataloguing of spectral lines. By devising a new system of notation, Russell and Saunders invented a means of keeping track of the complicated language of spectra.[103] Their work, building on decades of spectroscopy, helped to make nature's light intelligible.

---

[101]Bacon, *The New Organon and Related Writings*, 39.

[102]"Address as President of the Electrical Conference at Philadelphia" (1884), in *PPR*, 621.

[103]On the analogy between spectroscopy and the study of language, see Dörries, "Heinrich Kayser as Philologist of Physics."

CHAPTER

Astronomical Light in the West

The power of experiment and observation, after the manner of the Rowland school, developed especially strongly in Pasadena. There the engineering style in physical research had a home, and high–dispersion spectroscopy thrived. Caltech and Mount Wilson Observatory built on some of the achievements of the Hopkins school of physics, and gained stature by mastery of light. Lines of endeavor that the Hopkins school fostered were greatly furthered at the two institutions. John Anderson continued his instrument-building, and broadened well beyond spectroscopy. Using Hale's solar tower telescopes, investigators at the Observatory took up the painstaking task of checking and revising Rowland's solar tables. Henry Norris Russell, the Princeton astronomer who visited Mount Wilson frequently, set about extracting information on elemental abundances in the sun from the "Rowland intensities" of solar spectral lines. In this research, relations between lines in multiplets proved to be a valuable guide. The Mount Wilson emphasis on spectra, initiated by Hale, remained in force after Walter Adams became director of the observatory in 1923. Adams wrote, "The main problem of physical astronomy . . . is to interpret the inscriptions which we find contained in these stellar records, which we call the spectrum."[1] Technical advances continued to improve the prospects for high-dispersion spectroscopy, with attendant clarity, of the stars and other astronomical objects in addition to the sun.

---

[1]Walter S. Adams, "The Past Twenty Years of Physical Astronomy," *Science*, 67 (1928): 637-638.

Fine structure, a cardinal example of phenomena made evident in high-dispersion laboratory spectroscopy, received significant attention at Caltech. While Ira Bowen was still a graduate student of Robert Millikan, he and Millikan pursued hot-spark vacuum spectroscopy, which was no longer Theodore Lyman's exclusive domain. In 1920 they were able to push "three or four octaves farther into the ultraviolet" than any previous investigators.[2] Research on the spectra of ionized nitrogen, oxygen, and other elements led them into the study of fine structure. This research, the authors acknowledged, was made possible only by instruments manufactured at Mount Wilson by Anderson. Resolution and accuracy formed the authors' principal claim to importance; once again the engineering of observation underlay a bid for recognition.[3] Millikan and Bowen were able to extend doublet laws from x-ray spectra to optical spectra, and then to apply these laws to several elements in different states of ionization.[4] The subsequent work of Kent and Houston confirmed, in effect, the special status of Millikan's and Hale's institution in the determination of wavelengths.

Cutting-edge physics and astronomy, pursued in close proximity around Pasadena, were related in various ways. Both Millikan and his former colleague at Chicago, Michelson, saw a strong parallel between astronomy and laboratory spectroscopy.[5] The spectroscope was, to the atom, what the telescope was to the heavens—an instrument that extended the senses into previously impenetrable regions. Millikan wrote of an "astronomy of the atom," in which the spectroscope furnished "as exacting proof of the orbital theory of electronic motions as the telescope furnished a century earlier for the orbital theory of the motions of heavenly bodies." Millikan, the premiere physicist at Caltech, said that his study of the spectra of highly ionized atoms created new ways of "reading" the conditions existing in stars.[6] With the telescope and spectroscope combined, astronomy and spectroscopy were more than parallel: they were joined. This was especially clear when Bowen crossed into astronomy in 1927, finding that the "nebulium" lines, a mystery of astronomical spectra since the

---

[2]Robert A. Millikan, "The Astronomy of the Atom" (for Science Service, May 1924), in Millikan Papers Microfilm, Folder 59.48, 3.

[3]I.S. Bowen and Millikan, "The Fine Structure of the Nitrogen, Oxygen, and Fluorine Lines in the Extreme Ultra-Violet," *Philosophical Magazine*, 48 (1924): 259-260, 264.

[4]Ibid., 263-264.

[5]For Michelson, see "Recent Progress in Spectroscopic Methods," 362-363.

[6]Millikan, "Astronomy of the Atom," 2-3, 5. Some of Millikan's awareness surely reflected Hale's proselytizing. Hale wrote to a friend that Millikan, when being recruited for Caltech, was "beginning to see something of the bearing of the astronomical on the physical work, but none of us half realizes the real opportunity, in my opinion." Hale to Goodwin, 22 September 1921, GEHC.

preceding century, were emitted by ionized oxygen atoms in nebulae. This crossover, from the laboratory to the observatory, was just the kind of melding for which Hale had hoped in his institution–building, a hybridization of physics and astronomy. (Another crossover, between research and the engineering that supported it, was less visibly sought, but just as significant.) Henry Norris Russell felt sure of Bowen's identification from the start, because Bowen was, in Russell's words, "an excellent observer."[7]

On Hale's directive, some of Rowland's research program continued beyond the bounds of Rowland's school. The Baltimore physicist created instruments and trained physicists, but there was also a third lingering agent left by Rowland that impelled research and discovery well into the twentieth century: his solar wavelength tables, a potential rosetta stone with respect to the composition of the sun. As well as wavelengths, Rowland gave tentative elemental identifications of many lines, and estimates of the intensity of each line. Rowland and Hale, like others prior to the 1920s, believed the sun to be similar in composition to the Earth, but their work emphasized observation, not their own suppositions. Hale provided for the checking of Rowland's tables at Mount Wilson, and when Russell and colleagues approached the problem of solar composition in the 1920s, they were able to exploit "the wealth of data in Rowland's solar intensities."[8]

Rowland's own visual appraisal of the strength of each absorption line on his photographs constituted the original "Rowland intensity." In the late 1920s, the more precise concept of "equivalent width" arose to express line strengths more rigorously, in terms of the area darkened on a photographic negative. But Russell made his own calibration of Rowland intensities, and found them useful. From the widths and strengths of spectral lines, Russell tried to deduce solar abundances. His success derived from familiarity with both observational spectroscopy, American style, and emerging quantum conceptions. Making use of the new concept of multiplets, Russell compared differing line intensities within groups of lines, and found a way to derive relative numbers of emitting atoms in the source object. From this and other work, Russell thus confirmed what some astronomers, including himself, had considered but thought impossible: hydrogen was the most abundant element in the sun.[9] (Strong hydrogen lines in the solar spectrum had long been seen as anomalous.)

Solar tables did not remain unchanged over the decades. Russell maintained an interest as they were scrutinized and revised at Mount

---

[7]Russell to Paul W. Merrill, 22 September 1927, PWM.
[8]Hearnshaw, *The Analysis of Starlight*, 232.
[9]See DeVorkin and Kenat (II), 202-203, 208.

Wilson. In the same way that ruling engines evoked admiration from those who tried to duplicate them, Rowland's tables won praise from those who checked and revised them, even though some identifications changed.[10] Charles St. John took the greatest role in the revisions at Mount Wilson, where both the 60-foot and the 150-foot solar tower telescopes were used. The undertaking was consistent with the original mission of the observatory—solar research.

St. John also worked in a context of increasing internationalization, in which Rowland's values were compared with an emerging international system of wavelengths, fostered by international networks Hale had helped to create. The international wavelength system was anchored in 1907 by interferometric determinations of the red cadmium line. The adoption of this standard eventually required a correction (minus 0.212 Å) for constant error throughout Rowland's tables.[11] But for relative wavelengths, after correction, the tables were still fairly accurate. St. John called them "an instrument of precision in the hands of the solar and stellar investigator."[12]

Still, some relative errors existed. More observations were required, "to discriminate between that which is possible to retain and that which must pass." St. John reviewed attempts to find formulas to translate between the Rowland system and the new international system, but found that the differences varied unevenly. He maintained that "no factor of transformation nor any curve" could adequately describe the differences. The most significant discrepancies between St. John's measurements and Rowland's arose in regard to the separations of close pairs of lines. The discrepancies, St. John said, furnished "new grounds for not adopting any form of operator for transforming the wave-lengths" between the two major systems. The narrower the separation, St. John found, the greater the discrepancy between his and Rowland's measured separations.[13]

In refining Rowland's tables, St. John tried to reduce or eliminate any effects of a "personal equation" of the observer on measurements of close pairs. To do so, St. John borrowed some of Rowland's original solar spectrum plates from Johns Hopkins. Although these originals possessed a finer photographic grain than St. John's own plates, the latter were taken at a larger scale, which compensated. St. John found two sources of disagreement with his own wavelengths. First, Mount

---

[10]On the admiration, see Marcel Minnaert (1965), quoted in Hearnshaw, *The Analysis of Starlight*, 422.

[11]Charles E. St. John, "Revision of Rowland's Preliminary Tables of Solar Spectrum Wave-Lengths," *Proceedings of the National Academy of Sciences*, 13 (1927): 679.

[12]St. John, "The Situation in Regard to Rowland's Preliminary Table of Solar Spectrum Wave-Lengths," *Proceedings of the National Academy of Sciences*, 2 (1916): 226, 229.

[13]Ibid., 227-229.

Wilson measures of pair separations on the Rowland plates were less than Rowland's own measures on the same plates by an average of 0.0023 Å. Second, Mount Wilson measurements on Mount Wilson plates were less than Mount Wilson measurements of the Rowland plates by an average of 0.0022 Å. St. John attributed this latter effect to better photographic resolution on the Mount Wilson plates, "with a correspondingly lessened tendency toward overspacing." (The two effects summed to a difference of 0.0045 Å.)[14]

St. John found subjective "personal equation" differences that he attributed to effects of contrast in spectrum plates. He found that a greater contrast difference between a given line and the adjacent background on one side would make the line appear more intense on that side, and cause the observer to mark the position of the line further toward it. In St. John's words, the observer would tend to locate the maximum "nearer that edge of the line for which the contrast is greatest, i.e., nearer the free edge." In reaching his conclusions, St. John studied the personal equations of six Mount Wilson observers at work. He also compared spectrum photographs with plots of intensity versus wavelength as captured by a microphotometer. This device increased the scale of the spectrograms fifty times over, with a corresponding increase in accuracy. Mechanization thus helped to eliminate the personal equation.[15]

With greater line separation, the discrepancies between Rowland measures and Mount Wilson measures diminished, and St. John defended the overall integrity of Rowland's wavelength values. Some users assumed that accidental, relative errors in Rowland's wavelengths were significantly greater than 0.01 Å, but St. John found that, on average, relative errors were much less than 0.01 Å. Discrepancies between different sets of wavelength data, he said, represented "real differences in the behavior of lines" under different conditions.[16] Like Theodore Lyman studying ruling errors, St. John took an exacting Rowland creation, the solar tables, and demanded still greater accuracy. To do so, he used an Anderson grating (still from Johns Hopkins) in the 60-foot solar tower telescope, and a Michelson grating in the 150-foot solar tower telescope.[17]

In the study of Rowland's errors there was potential for gain. Research at Mount Wilson pushed accuracy to further decimal places, and made the record less ambiguous; close pairs were delineated more

---

[14]St. John and L.W. Ware, "The Accuracy Obtainable in the Measured Separation of Close Solar Lines; Systematic Errors in the Rowland Table for Such Lines," *ApJ*, 44, (1916): 32-33.

[15]Ibid., 17-18, 20-21, 34-36, 38.

[16]St. John, "The Situation," 229.

[17]St. John and Ware, "The Accuracy Obtainable," 16.

carefully. Rowland's original plates provided points of reference. Quite unexpectedly, St. John's revision led him into the position of providing observational support for general relativity. A skeptic in 1917, St. John favored general relativity by 1923, when he was convinced that instances of gravitational redshift, as specified mathematically, actually occurred in the solar spectrum, even though the effects in question amounted to less than 0.01 Å.[18] (Separately, Hopkins-school alumnus Joseph Moore, working at Lick Observatory in 1928, confirmed gravitational redshift in the case of the star Sirius B, supporting an earlier measurement by Walter Adams at Mount Wilson, although in the 1960s the work of both observers was discredited by new measurements, with a different redshift value.)[19]

A few years after St. John's re-examination of the Johns Hopkins solar plates, William Meggers found Rowland's material legacy in great disarray. Like St. John, Meggers wished to see the original solar plates. He entered the Baltimore building where they were kept, and found them "thrown in careless disorder on the floor of the attic." For Meggers, who regarded Rowland's researches "with profound reverence," the condition was shocking. He wrote to Joseph Ames "to call attention to the negligence and disrespect shown to the intrinsically and historically valuable photographic records of Professor Rowland's work." It was the duty of the university, he felt, "to protect and cherish" these plates.[20] Thus Meggers displayed his desire to conserve the Johns Hopkins heritage.

But the discipline of spectroscopy had what it needed, regardless of the condition of the original plates. Rowland's tables were long since published, and many copies of his spectrum maps had been sold. The plates had been transformed into inscriptions of wavelength, and these inscriptions were in turn refined elsewhere. As for Meggers, his dialogue with Joseph Ames at Johns Hopkins turned from dated plates to working gratings. He was most interested in the log books recording past operations of the ruling engines, from which he hoped to extract the causes of ghosts from specific instruments.[21] Again it was Rowland's errors, at the limit of his powers, that constituted the research frontier.

Meggers also held an interest in the solar spectrum, and mapped lines in the red and infrared at about the same scale as Rowland's

---

[18]*Klaus Hentschel, "The Conversion of St. John," Science in Context*, 6, 1 (1993): 137-194. Separately, Hentschel also suggests that Rowland himself unwittingly detected redshifts three decades earlier, but largely dismissed them as flawed observations. Hentschel, "The Discovery of the Redshift of Solar Fraunhofer Lines by Rowland and Jewell in Baltimore around 1890," 219-277.

[19]Norriss Hetherington concludes that Adams and Moore were "eluding the constraints of objectivity to find what they expected to find, even when it didn't exist." *Science and Objectivity*, 70-72.

[20]Meggers to Ames, 28 February 1919, WFM.

[21]Meggers to Ames, 1 April 1919, WFM.

photographic maps, finding that most of the absorption lines in this region of the spectrum were terrestrial (atmospheric) in origin.[22] He remained in close contact with colleagues in Pasadena, including Paul Merrill, who had worked at the National Bureau of Standards from 1916 to 1919. During his first year at the observatory, in 1919, Merrill thought it would be appropriate for Meggers at the Bureau to proceed with a solar map that would emphasize chemical identifications. He wrote, "The solar spectrum doesn't belong to Mt. Wilson so you are free to do whatever you wish with it so long as you don't injure it any." Merrill suggested, however, that Meggers leave wavelength determinations of high accuracy to Mount Wilson.[23] The superior equipment there, especially the solar tower telescopes, probably gave the observatory an unbeatable advantage.

Merrill had left the Bureau for Mount Wilson to carry out his own ideas "for systematic work in the observation and interpretation of the less understood types of stellar spectra." Stellar spectra had been taken in great abundance by 1919, and Merrill's emphasis was consistent with Hale's desires to extend high-dispersion studies from the sun to the stars. Merrill wrote Hale, "I would put the emphasis here on the word systematic for no program of that kind involving high dispersion and all regions of the spectrum has been undertaken. Recent observations of these objects have been undertaken largely fragmentary [sic] and at random."[24]

From Mount Wilson, Merrill reported back to Meggers on the variety of research being carried out. Robert Millikan had started a journal club at Caltech. John Anderson at the Observatory was observing electrically exploded wires, to obtain high-temperature spectra. Merrill wrote, "He sends a terrific jolt from a large condenser through a fine wire and blows it up. The spectrum of this contains as beautiful absorption lines as you would care to see." And there were other attempts to reproduce in the laboratory the extreme heat of the sun and stars. Arthur King studied absorption spectra from matter heated in an electric furnace. He used an arc with a current of up to 1,000 amperes, in vacuum, to heat samples in tubes. As Merrill described it,

> They would only last about thirty seconds or so as the tubes burn away so fast. Designs are now in the shop for a furnace where one tube can be fed in from the outside so that an arc can be maintained for some time.

---

[22]Meggers to Lewis E. Jewell, Eastman Kodak Company, 8 January 1919; Frank Schlesinger, Allegheny Observatory, to Meggers, 10 March 1920; Meggers to Schlesinger, 13 March 1920, WFM.

[23]Paul W. Merrill to Meggers, 23 September 1919, WFM.

[24]Paul W. Merrill to George Ellery Hale, 19 August 1918, PWM.

Anderson, going further, planned to create a large solenoid that would produce a magnetic field big enough to envelop a small furnace. With this device, the Zeeman effect in heated matter in the laboratory would be susceptible to analysis and comparison with the same effect in the sun and stars.[25] (King used concave gratings in at least some of his investigations.)[26]

Merrill himself studied stellar radial velocities, hoping that a new prism spectrograph would advance this work. In the early 1920s, he began experimenting with concave gratings attached to the 100-inch telescope, then still the largest in the world.[27] From Merrill's point of view, research at Mount Wilson overwhelmingly favored spectra: "Nobody here makes a business of astronomy of position." Merrill added this to his reasons why Mount Wilson astronomers were not, in 1920, actively pursuing research on the deflection of light rays (a change in apparent stellar position) in the gravitational field of the sun, predicted by general relativity.[28] Back east, Meggers tried to find a way to record any deflection of rays of starlight near the massive sun without the need to wait for an eclipse to eliminate extraneous light from the terrestrial atmosphere. Meggers suggested to another colleague that, since "skylight is almost entirely blue and violet," it might be extinguished by a filter admitting only infrared light. With haze from the atmosphere eliminated, he suggested, stars near the sun might even be photographed in the daytime. Meggers encountered skepticism from his colleagues, however, and the project was apparently never attempted. Still, Meggers remained interested in attempts in Germany to measure shifts in wavelength predicted by general relativity, and asked Heinrich Kayser whether success had been achieved there: "I will be glad to know that this is true since Mt. Wilson astronomers are very much opposed to the 'Einstein foolishness' as some of them call it."[29] (Merrill and others held this skeptical attitude even after the 1919 British confirmation of

---

[25]Merrill to Meggers, 1 March 1920, WFM.

[26]Arthur S. King, "Electric Furnace Experiments Involving Ionization Phenomena," *ApJ*, 55 (1922): 381. Babcock notes that King used Anderson gratings. Babcock, "Diffraction Gratings at the Mount Wilson Observatory," 158.

[27]Merrill to Meggers, 1 March 1920, 2 March 1922, 13 December 1922, WFM.

[28]Merrill wrote,

> Nobody here is very enthusiastic about the Einstein stuff. I have heard it referred to as an "accursed theory." Some of the coolest Englishmen even have no use for it. There is not much violent opposition to it here—the attitude is one of watchful waiting. There are dozens of other things which are more worth working on.
>
> *Merrill to Meggers, 1 March 1920, WFM.*

[29]Meggers to Frank Schlesinger, Allegheny Observatory, 9 January 1920; Schlesinger to Meggers, 17 January 1920; Keivin Burns to Meggers, 6 February 1920; Meggers to Heinrich Kayser, 28 April 1920, WFM.

the deflection of starlight, but a decade later the situation was much different, and a personal visit by Einstein met with general acclaim. At a dinner in Einstein's honor, Millikan proudly listed 25 "Einstein collaborators" at Caltech and Mount Wilson, including himself, Bowen, St. John, Walter Adams, Edwin Hubble, Michelson, Robert Oppenheimer, and King, but not Merrill.)[30]

Solar data from Mount Wilson was important for the contradiction of expectations, whether those of Hale, St. John, or others. Into the 1920s, astronomers proceeded on the assumption that the sun and stars were similar in elemental composition to the Earth. This increased the surprise when observation and cross-comparison with laboratory results gradually led to the conclusion that hydrogen dominated the solar atmosphere. The finding arose through complex measurement and calculation, but it was through efforts launched by Hale, first and last a solar physicist, that much of the data was accumulated. The determination of solar abundances (and thus, approximately, stellar abundances) should take a historiographical place alongside the discovery of the size of galaxies, and the detection of the cosmological redshift, as a major astronomical revolution of the 1920s.

The discovery was not purely observational, but access to Mount Wilson observations was critical to Henry Norris Russell, the astronomer who effected the change in consensus. (Cecilia Payne accepted a high hydrogen abundance in the mid-1920s, but maintained her position quietly because Russell and other leaders at first could not agree with her.) Russell was a frequent research associate at Mount Wilson, especially in summers and winters. His own assistant, Charlotte Moore, made an extended stay at the observatory in 1926 and 1927, to assist in a revision of the Rowland tables in those years. According to DeVorkin and Kenat, access to the most advanced spectral observations was an essential ingredient of Russell's work.[31] Of course, others had access but not Russell's insight.

Russell knew that Meggers and others saw a relationship between the strength of spectral lines and the corresponding amount of each element present in the source of light. He knew that Rowland had used two measures of abundance: the intensities of lines (which placed calcium as the most abundant element in the sun), and the number of lines (which placed iron first). Russell had some familiarity with emerging quantum ideas, especially on multiplets. And, importantly, he had direct access to observational data at Mount Wilson.

---

[30]Millikan, "The Reason and the Results of Dr. Einstein's Visit to the California Institute of Technology," *Science*, 73 (1931): 381.

[31]DeVorkin and Kenat (II), 216.

This access was especially advantageous because these observations were rarely challenged.[32]

Russell had initially hoped that spectroscopic data would be mobile enough to reach his home university. He wrote that Theodore Lyman and another colleague

> both heartily approve an idea of mine for making Princeton a center, for the time being, for collecting data on these new regularities. We have an excellent computer [a person who computed] and stenographer here, who have some spare time, and I should enjoy heartily the general oversight of the material.[33]

However, at least in the case of solar spectra, Pasadena remained the center, to which Russell and his assistant, Moore, had to travel.

As Russell developed his understanding of the atom, he "compared it in detail to the wealth of spectroscopic evidence that A.S. King [of the furnace experiments] had been carefully gathering over the past decade." The agreement with observation excited both Russell and Meggers. Russell collaborated with King, Meggers, St. John, and others, and had the constant help of Moore. In 1924 and 1925, Russell had Moore gather information on the characteristics of spectra from Mount Wilson and those from Rowland's own solar atlas, and concluded that abundance was the main cause of line broadening. (Other possible causes included, for example, pressure and the Stark effect.) He pressed for a revision of the Rowland atlas, which might provide a better guide. Rowland intensities were "the only large-scale store of information then available."[34]

In late 1925, St. John and Walter Adams made plans to revise the Rowland atlas. Adams asked Russell to loan Moore for the project, and she arrived in January, to work with St. John, Harold D. Babcock, and others. (Babcock had worked at the Bureau of Standards from 1906 to 1909, and later took charge of the Mount Wilson ruling engine.) The revision took two years. At the conclusion of the project, Russell, Adams, and Moore presented a calibration of Rowland intensities, based on revised measurements and on Russell's knowledge of relationships between lines in multiplets. They found a logarithmic relationship between Rowland intensities and the relative, not absolute, number of emitting atoms in the sun. A small difference in line strength corresponded to a large difference in abundance. A large difference in line strength, from very weak to very strong, might correspond to an increase by a factor of one million.[35]

---

[32]Ibid., 181, 186, 216.

[33]Russell to Meggers, 6 February 1924, WFM.

[34]DeVorkin and Kenat (II), 196-197, 199-200, 208.

[35]DeVorkin and Kenat (II), 201-203. Russell, Walter S. Adams, and Charlotte Moore, "A Calibration of Rowland's Scale of Intensities for Solar Lines," *ApJ*, 68 (1928): 8.

In the late 1920s, a growing number of astronomers suspected a high hydrogen content, and Russell was not alone in trying to extract the significance of line strengths. The work of Albrecht Unsöld in Germany, for example, enabled Russell to obtain true, not simply relative, abundances. But, with the observational groundwork laid, Russell worked intensively at Mount Wilson in the winter of 1928-29, and came to accept the modern view, that the Earth and Sun are fundamentally dissimilar in composition. Yet the empirical evidence was still as slender as the spectral lines, and the new view was founded on "the vague physical meaning of the intensities themselves, which were only very rough estimates of photographic line width."[36] Later, more exact, relations between line width and abundance were developed mostly in Europe.[37]

The Mount Wilson revision of the Rowland tables, published in 1928, added new categories of information on spectral lines from the sun. Arthur King had by then classified many laboratory lines on the basis of the temperature at which they were produced. Class I lines, at the lowest temperature, occurred in flame spectra. Class II and III lines arose at higher temperature, in the electric arc. And Class IV and V lines arose in the spark, at very high temperature. Generally, higher temperatures corresponded to greater states of ionization.[38] The revised Rowland tables included this temperature classification for each solar line, if it was known from the laboratory. The new tables also classified the behavior of some spectral lines when the matter responsible for the lines was under varying pressures, and gave separate information on the intensities of lines that occurred in sunspots as well as those that occurred in the solar disk. Finally, the tables gave the atomic excitation potential necessary for those lines that had been classified according to spectral series or multiplet group. This data was added to the original three columns: wavelength, identification by element, and intensity.

The tables built on more than ten years of observation at Mount Wilson, including reconciliation with international primary and secondary wavelength standards as determined by interferometer. (These wavelength standards, too, were an offshoot of Hale's endeavors, because they were established by the International Union for Cooperation in Solar Research, which Hale had helped to create.)[39] As for

---

[36]DeVorkin and Kenat (II), 206-210.0

[37]Hearnshaw, *The Analysis of Starlight*, 244.

[38]Donald H. Menzel, "History of Astronomical Spectroscopy II: Quantitative Chemical Analysis and the Structure of the Solar Atmosphere," *Annals of the New York Academy of Sciences*, 198 (1972): 236.

[39]St. John, "Revision of Rowland's Preliminary Tables," 679-680, 682-683.

the actual compilation of the tables, much of the credit belonged to Charlotte Moore.[40]

Hale's work of organization in Southern California coalesced around the gathering and dispersion of sunlight, and it quickly ramified into other areas. It retained the engineering approach to acquiring observations. Hale's concern lay as much with the technical questions of how, materially, to accumulate knowledge, as with the knowledge itself. This style is evident in the work of John Anderson. Numerous solar and laboratory studies at Mount Wilson recorded some debt to Anderson's ruling operation. St. John had the benefit of an Anderson grating even while Anderson was still being recruited for a permanent position.[41] By the time Anderson came to Pasadena, the ruling engine was working smoothly,[42] and his attentions were not limited to that machine.

Anderson pushed the limits of performance in several areas. In his investigations of the Stark effect, he used high voltages to create a source bright enough to furnish spectra of higher dispersion than previously attained, with correspondingly finer detail. With a Rowland grating, he recorded seventy-four lines of chromium, accurate to 0.01 Å, as affected by an electric field.[43] Even where electric fields were not of primary interest, Anderson still employed high voltages, to re-create stellar conditions in the laboratory, sending extreme electric currents through fine wires to make them burst into ephemeral cylinders of hot, luminous vapor. (At Caltech in the 1920s, Millikan and Bowen used hot sparks to strip electrons from atoms and study the spectra of elements in an ionized state, with limited numbers of electrons in the outermost orbit.)[44]

Spectrographs were of course essential to the laboratory studies. In the Mount Wilson ruling operation, as ever, extreme mechanical stability was essential; even slight Earth tremors might lead to flaws in gratings. With seismologist Harry O. Wood, Anderson developed a torsion seismometer to keep track of potential flaws. He expected only to record local earthquakes, but in fact the device was sensitive enough to register the initial phases of earthquakes as far distant as Japan.[45] The component of the seismometer subject to torsion was a tungsten wire 0.02 millimeters in width. A small plane mirror,

---

[40]Menzel, "The History of Astronomical Spectroscopy II," 242.

[41]Hale to Anderson, 22 December 1915; Anderson to Hale, 6 January 1916, GEHM.

[42]Anderson to Hale, 7 July 1916, GEHM.

[43]Anderson, "A Method of Investigating the Stark Effect for Metals, with Results for Chromium," *ApJ*, 46 (1917): 108-109, 116.

[44]Bowen and Millikan, "The Series Spectra of the Stripped Atoms of Phosphorus (P_5), Sulphur (S_6), and Chlorine (Cl_7)," *The Physical Review*, 25 (1925): 295-305.

[45]Goodstein, *Millikan's School*, 135.

attached, rotated with the wire if it twisted, and consequently deflected light, providing a measure of seismic disturbance.[46] Thus, in the pursuit of spectroscopic accuracy, Anderson fostered, as a byproduct, a better way to observe the Earth.

Anderson also saw great potential in the application of the interferometer to astronomy. It showed such promise that Hale asked Anderson to explore possible uses.[47] After working out plans for constructing an interferometer with Francis Pease, who was a mechanical engineer by training, Anderson chose to observe the double star Capella, with an orbit known from spectroscopic data. Using the 100-inch telescope with interferometer, he found the angular separation of the two components to be 0.05249 seconds of arc, and the corresponding distance of separation to be 130,924,000 kilometers. (He was also able to observe the star in daylight, but found that the accuracy was not as good.)[48] Soon after Anderson's work appeared, others published a study of the diameter of Alpha Orionis.[49] Anderson was thus a facilitator, adapting a difficult new technique and fostering implementation in practice; his study of Capella was a demonstration.

Anderson's success in the role of supporting others' observations, typical of the Hopkins school, probably contributed to his selection in 1928 as executive officer of the largest project that Hale undertook—the planning and construction of a 200-inch telescope. The project was motivated in part by Hale's desire to realize a former goal of the Hopkins school, the acquisition of large-scale spectra of the stars. Hale wrote: "With a larger telescope, we could push the dispersion to the point attained by Rowland in his classic studies of the solar spectrum."[50] Anderson worked up a cost estimate, discussed an astrophysical laboratory for Caltech to accompany the new telescope, directed a site search, and explored the inherent limitations on large telescopes. The site search was crucial because ambient conditions critically affected telescope performance, and Hale asked Anderson to compare the brightness of the night sky at various possible sites. (Light pollution, a twentieth-century vexation, was already a concern.) At the site eventually chosen, Mount

---

[46]Anderson and H.O. Wood, "A Torsion Seismometer," *Journal of the Optical Society of America and Review of Scientific Instruments*, 8 (1924): 818, 822.

[47]Anderson, "Application of Michelson's Interferometer Method to the Measurement of Close Double Stars," *ApJ*, 51 (1920): 263 & note, 264.

[48]Ibid., 264, 268.

[49]Michelson and Pease, "Measurement of the Diameter of Alpha Orionis with the Interferometer," *ApJ*, 53 (1921): 249-259.

[50]Hale, "The Possibilities of Large Telescopes," *Harper's*, 156 (1928): 644.

Palomar, meteorological conditions were monitored by recording instruments for five years.[51]

Having analyzed the image resolution attainable at the planned observatory by means of an interferometer attached to a large telescope, Anderson also explored, soon after his appointment, the limits on image resolution imposed by the surrounding atmosphere. A large telescope had little advantage if seeing conditions were poor. The atmosphere, Anderson reasoned, made a star image ten to twenty times larger than it would otherwise be, and one percent, or less, as bright. It made the image into "a boiling patch of light."[52] He even considered the effect of wind pressure on the planned telescope tube, and concluded that the maximum momentary displacement of the tube inside the dome would be 1.5 thousandths of an inch, or 0.5 arc seconds in the focal plane, in an extreme gust.[53]

Creation of the rough disk for the mirror and transportation of the disk from New York to California caused a public sensation, but actual figuring of the mirror in Pasadena posed an unprecedented challenge. Few opticians could approach the problem, so Anderson "assembled a crew of untrained men and taught them the necessary techniques." Along with the telescope itself, Anderson saw the need for many auxiliary instruments to handle the light collected by the large mirror, and supervised work on these instruments. A colleague later recalled, "These spectrographs were tried first on the 100-inch telescope and made possible many new fields of spectroscopic study including the observations which led to the concept of the expanding universe."[54] Anderson was an enabler; the record of his publications does not reveal the full measure of assistance he afforded to astronomy and astrophysics.

While Anderson managed Hale's grandest venture into big science, continued efforts took place to realize Hale's early goal of obtaining high-dispersion stellar spectra that might reveal as much about the stars as the Rowland tables revealed about the sun. American efforts with gratings up to the 1920s were spotty. Crew (using a concave grating on loan from John Brashear) and Keeler (with a Rowland plane instrument) made separate attempts in the 1890s. (Rowland himself did not attempt stellar spectra.) Hale and Adams managed an extended exposure of Arcturus in 1905 with a plane grating; W.W. Campbell at Lick Observatory observed Mars in 1910; Adams pursued

---

[51]Hale, "The Astrophysical Observatory of the California Institute of Technology," *ApJ*, 82 (1935): 114, 126. Hale to Anderson, 20 July, 23 July, 19 November 1928; Anderson to Hale, 20 July, 23 July 1928, 20 May 1929, 27 March 1930, GEHM.

[52]Anderson, "Larger Telescopes," in Anderson Correspondence, 1929, GEHM.

[53]Anderson to Hale, 3 February 1933, GEHM.

[54]Bowen, "John August Anderson," 10-11.

stellar spectra with the 60-inch telescope and a concave grating in 1913; and J.S. Plaskett in Ottawa concluded in 1914 that plane grating spectra, although weak, held promise in certain applications.[55]

In the event, "[c]oncave gratings never did catch on for stellar spectroscopy,"[56] but the plane version gained ground. Although, as of 1931, gratings were "not . . . extensively employed in stellar spectrographs,"[57] the situation changed rapidly afterward. A pioneer user was Paul Merrill, who came to Mount Wilson specifically for high-dispersion studies, and pursued them throughout the 1920s. Toward the conclusion of the First World War, Merrill developed photographic emulsions sensitive to long wavelengths, believing them to have value both for military and for astronomical work. At Mount Wilson after the war, Merrill worked on a system by which the most appropriate photographic chemicals could be adopted for each of the various kinds of astronomical observation. Then, in 1922, despite the general preference for prisms at the time, Merrill began to experiment with gratings attached to the 100-inch telescope.[58]

Merrill found that the grating had the following advantages: "(1) the exposure times are shorter; (2) a longer range of the spectrum is obtainable (with full illumination) on one plate; (3) the dispersion is normal; (4) a change from one spectral region to another is more easily made; (5) the spectrograph is smaller and less expensive." Merrill obtained eleven stellar spectrograms in 1922, with a dispersion of 33 Å per millimeter. Because the focal plane of the concave grating was curved, Merrill had to use film rather than glass plates to record the spectrum. Departure from mechanical stability posed a problem for Merrill, as it had for previous investigators. However, Merrill was able to obtain a spectrum of Arcturus in one hour, where Hale and Adams had needed twenty-four hours, cumulatively over five nights. (Their dispersion was higher, 4.3 Å/mm.)[59]

Merrill then turned to a plane instrument, which he called a Jacomini grating (for the engine operator at Pasadena, rather than the supervisor, Anderson). With a colleague he obtained 120 spectrograms over two years. The focal plane was not curved, so he could employ flat plates. Jacomini had shaped the grooves to concentrate light in the first-order spectrum, for greater brightness. However, flexure during expo-

---

[55]Osterbrock, "Failure and Success: Two Early Experiments with Concave Gratings in Stellar Spectroscopy," 119, 121, 125. Paul W. Merrill, "A Plane-Grating Spectrograph for the Red and Infra-Red Regions of Stellar Spectra," *ApJ*, 74 (1931): 189-190.

[56]Osterbrock, "Failure and Success," 127.

[57]Merrill, "A Plane-Grating Spectrograph," 189.

[58]Merrill to Hale, 19 August 1918; Merrill, "Photographic Testing at Mount Wilson Observatory," Folder 2, Box 1, PWM. Merrill, "A Plane-Grating Spectrograph," 188-189.

[59]Merrill, "A Plane-Grating Spectrograph," 189-190. Osterbrock, "Failure and Success," 125.

sure, which Merrill attributed to poor rigidity of the mounting, caused displacement of spectral lines. Still, Merrill was satisfied that he had "demonstrated the usefulness and efficiency" of the instrument.[60]

Plans then centered on the creation of a mounting that would eliminate flexure. Following a compact design, the main casting for the new mounting had a v-shape, with the grating to be placed at the vertex. The plane instrument used a specially-built collimator, designed by W.B. Rayton of Bausch and Lomb. The highest dispersion achieved with the apparatus was 34 Å/mm. An outer cover insulated the apparatus thermally, and contained both a thermostat and a thermometer. The mounting made provisions for positioning a vacuum tube of neon, argon, or helium, to provide a comparison spectrum. The new mounting, with mechanical rigidity and thermal stability, virtually eliminated the problem of flexure, making possible exposures of three to six hours. Merrill felt that credit was due, not just to one individual, but to the entire observatory shop.[61]

With the new apparatus, Merrill and other investigators were able to photograph some stellar lines which, Merrill felt, called for greater study. There were lines which Merrill thought might be interstellar in origin; detailed behavior of lines in the Balmer series held interest; stellar radial velocities of many stars became easier to obtain (higher dispersion meant greater accuracy, necessary in velocity studies); and new lines still requiring identification were recorded. In one case, Merrill faced the dilemma of deciding whether a bright patch of spectrum was an emission line or just "a narrow section of the continuous spectrum" lying between two dark absorption bands. In another case, a nebular line at 7325 Å was photographed, and the line structure confirmed Ira Bowen's identification of "nebulium" as ionized oxygen at high temperature and low density.[62]

Merrill soon found other lines, in stellar spectra, that he identified as "forbidden" lines from ionized iron. This gave him something to offer his former colleagues at the Bureau of Standards: "I have just found that many of the 'forbidden' lines of Fe II are strong in certain stars, although not observed in the laboratory. Probably the explanation is connected with a great mean free time, and the facts may have a general bearing on theories of spectroscopic transitions. If any at the Bureau are interested I would be glad to give you the new stellar dope on Fe II." In return, Merrill hoped for know-how from the Bureau, still a valuable resource, on emulsion-making and

---

[60]Merrill, "A Plane-Grating Spectrograph," 190-191.
[61]Ibid., 191, 194-195.
[62]Ibid., 195-199.

optical glass, as well as the latest combination rules for spectral series terms.[63]

In 1929, Merrill received a new grating from Jacomini that concentrated light in the first-order spectrum, offering better brightness.
With the new Jacomini instrument included, Merrill's spectrograph
for the 100-inch telescope then represented a milestone. Hearnshaw
states: "This was the first modern and successful grating spectrograph in regular use in stellar astronomy."[64] (After Anderson moved
from the ruling operation to the 200-inch telescope project in 1928,
Harold D. Babcock took over the former, and supervised production
until 1947.)[65]

Although Jacomini began in the 1920s to shape grooves to concentrate light in a given spectral order, in the 1930s Robert Wood at
Johns Hopkins produced "blazed" gratings in which control of groove
form handled light even more efficiently and furnished significantly
brighter spectra. Wood supplied one of these new gratings to spectroscopists at Harvard. Walter Adams at Mount Wilson learned of
Wood's success, and inquired about availability, informing Wood that
these products would be very useful at the coudé focus of the 100-inch
telescope.[66] (For the coudé focus, light was passed down one axis of
rotation of the telescope, to a fixed location where heavy instruments
requiring great mechanical and thermal stability could be accommo-
dated.) Adams also knew about Wood instruments from Mount Wilson astronomer Theodore Dunham, who reported to him on Wood's
system of obtaining Pyrex discs—having them figured, and having
them coated with aluminum, which placed less demands on the diamond in the ruling engine.[67]

Adams believed that the Mount Wilson shop could match
Wood's output, but only after a long delay. Hence, to obtain an instrument sooner, he petitioned Wood, who acted as his own agent
and named a price of $700 for an instrument, due to the difficulties of
production. Adams was still anxious to obtain one, to handle the
large four-inch-wide beam of light gathered by the large telescope.
Stellar spectroscopy required it. And Adams felt the dispersion of
Wood's existing instruments would be sufficient, without the need
to push for more. Adams informed Wood, "There are only one or two
stars in the sky which are bright enough for higher dispersion, so a

---

[63]Merrill to Meggers, 29 November 1927, PWM.
[64]Hearnshaw, *The Analysis of Starlight*, 12.
[65]Babcock, "Diffraction Gratings at the Mount Wilson Observatory," 157.
[66]Walter S. Adams to Wood, 24 August 1935, Walter S. Adams Papers (WSA), Carnegie Observatories Collection, Huntington Library, San Marino, California.
[67]Theodore Dunham to Adams, 1 April 1936, WSA.

bright first order grating is just what we need." Adams even arranged for the necessary Pyrex disc to be manufactured at the Mount Wilson shops.[68]

The acquisition of a Wood grating would help maintain the standing of Mount Wilson as a preferred research installation. Dunham, who had studied with Henry Norris Russell, had considered working elsewhere, but Adams countered in part by assuring him that Mount Wilson offered "a better chance for very fine high dispersion work on stellar spectra." The acquisition of a Wood grating was a way of fulfilling the promise. (Merrill, too, had been lured to Mount Wilson by the prospect of high-dispersion work.) The value of the instrument was great enough that Adams arranged for Dunham, traveling in England and also visiting Harvard, to receive it personally in Baltimore, on his way back to Mount Wilson. After Wood gave the instrument to Dunham, he wrote to Adams that it was "probably the most remarkable grating that has ever been ruled." Adams agreed, and noted the "amazing amount of light in the first order." Dunham was already aware of the potential. He later recalled, "That was when we decided to drop [prism-maker] Zeiss and go to gratings."[69]

The problem of handling light efficiently was not limited to diffraction instruments. The absorption of light by lenses was still significant, and the invention of the Schmidt camera in Germany in the early 1930s made possible the handling of light solely by reflection, without lenses, minimizing absorption. The brightness of the Schmidt camera opened up new possibilities, and, coupled with blazed gratings, offered much more efficient handling of light. According to Hearnshaw, "The use of the Wood grating and Schmidt cameras revolutionised high dispersion stellar spectroscopy from 1936." Gratings soon dominated stellar spectroscopy, while prisms declined in favor.[70] Looking back, Charles Fehrenbach has identified three historical periods in spectrographic work in the twentieth century in which different instruments prevailed: prism spectrographs used on large refractors (1898-1910); prism spectrographs used on special-purpose reflecting telescopes (1910-1930); and large grating spectrographs used

---

[68]Adams to Wood, 24 August 1935, 4 September 1935; Wood to Adams, 1 September 1935; Adams to Dunham, 4 April 1936, 13 May 1936, WSA.

[69]Adams to Wood, 4 August 1936; Adams to Dunham, 28 April 1936; Wood to Adams, 14 October 1936; Adams to Wood, 21 October 1936; Dunham to Adams, 1 April 1936, WSA. Interview with Dunham, 1977-1978, Sources for History of Modern Astrophysics, American Institute of Physics, 54. Dunham remembered: "I came back, on a Pullman sleeper, and I had the grating under my pillow. Didn't want it to be picked up by a porter and put off in St. Louis somewhere!" Ibid., 52.

[70]Hearnshaw, *The Analysis of Starlight*, 14-17.

with Schmidt cameras (from 1931). Innovations leading to brighter spectra have meant that "practically all spectrographs made since 1948 employ diffraction gratings."[71]

Like Bowen and Merrill, Dunham soon discovered "forbidden" behavior of electrons lying beyond the terrestrial environment. Like them, he studied emissions from ionized atoms, and found that nature sometimes defied the selection rules that physicists had developed to "forbid" transitions which were not, as a rule, observed in the laboratory. The energies resulting from forbidden transitions, and thus the wavelengths of corresponding emissions, could be calculated, and in some instances matched up with observed spectral lines, as in the cases of Bowen and Merrill. Observational spectroscopy was thus a source of novelty.

Dunham worked in collaboration with Adams on the 100-inch telescope, using a Wood grating and a Schmidt camera. With a dispersion of 10.2 Å/mm, Dunham measured stellar lines to the nearest hundredth of an angstrom—approaching the level of accuracy used by Rowland in his initial measurements of solar absorption-line wavelengths. In the spectra of five stars, Dunham found two sharp ultraviolet lines, "the appearance of which suggests an interstellar origin." He then searched spectral tables for comparable laboratory lines, and found only two, both resulting from ionized titanium (Ti II). Laboratory results suggested that other, fainter titanium lines might be found, and a longer exposure in the observatory successfully recorded them, confirming Dunham's identification.[72] Reasoning about the atomic transitions responsible for the lines led Dunham to calculate an upper limit on the density of matter in interstellar space—approximately 10^{-16} to 10^{-18} grams per cubic centimeter. Separate evidence suggested a much lower maximum density, so Dunham concluded that interstellar lines from other elements should be sought, as well, to improve the estimate. In general, Dunham saw good prospects for the detection of other interstellar lines.[73]

Even more than Dunham, Adams made clear the advantages of the spectroscopic innovations at Mount Wilson. Development of the Schmidt camera and of Wood's blazed gratings led the staff to completely reconstruct the spectrograph for the 100-inch telescope. The new spectrograph was isolated even from the concrete pier which had anchored previous designs, but which suffered slight strains. An

---

[71]Charles Fehrenbach, "Twentieth-Century Instrumentation," in Owen Gingerich, ed., *Astrophysics and Twentieth-Century Astronomy to 1950: Part A* (New York: Cambridge University Press, 1984), vol. 4 of *The General History of Astronomy*, ed. Michael Hoskin, 177, 179.

[72]Dunham, "Forbidden Transition in the Spectrum of Interstellar Ionized Titanium," *Nature*, 139 (1937): 246-247.

[73]Ibid., 247.

"image-slicer," invented by Ira Bowen, made certain that more light from the round telescope image passed through the narrow slit which served as a gate for the spectrograph. The use of reflection in the Schmidt camera and grating, as well as the telescope, meant that there was "no glass or quartz in the entire optical system between the star and the photographic plate." (Six reflections took place.) This innovation, eliminating most light absorption, made possible much shorter exposure times, especially in the blue end of the spectrum; in the red end, the reflections led to significant light loss. Adams reported a dispersion as good as 2.9 Å/mm, and a resolution sufficient to separate lines differing only by about 0.1 Å in wavelength. The new arrangement could record high-dispersion spectra of stars as faint as magnitude 6.0.[74] This was the realization of Hale's dream.

By 1941, Mount Wilson astronomers had, Adams reported, used the new spectrographic arrangement in several investigations. One involved measurements of stellar radial velocity in an attempt to derive the solar parallax (motion of the sun through space). Observers chose as a reference star Arcturus (a Bootis) because lines in the spectrum remained "sharp and narrow under the highest dispersion." Therefore precise measures of line displacement due to radial motion could be made. Altogether, the staff took 37 spectrograms with the new grating, and then chose 25 lines for measurement on each exposure. Wavelengths were then compared to values in the revised Rowland table, and to comparison lines from the iron arc spectrum in the laboratory. By means of methods of calculation developed elsewhere, a velocity of 5.621 kilometers per second was obtained, with a probable error of less than 0.005 km/sec. Adams attributed the error to the nature of the iron comparison arc, and possible temperature fluctuations in the spectrograph. Assuming that errors could be reduced, Adams believed that radial velocity measurements offered a cost-effective approach to obtaining the solar parallax to high accuracy.[75]

In general, studies at high dispersion made possible fine distinctions in the treatment of spectral lines. Studies of possible water-vapor lines in the spectrum of Mars used the radial velocity of that planet and the corresponding shift of lines to distinguish between terrestrial water-vapor lines, from the atmosphere, and any possible Martian lines. The average shift detected was only 0.006 Å, below the level of significance, after potential error in measuring the somewhat diffuse lines was considered. But the negative result made possible the conclusion that, "if water vapor lines are present in the spectrum of the

---

[74]Adams, "Some Results of the Coude Spectrograph of the Mount Wilson Observatory," *ApJ*, 93 (1941): 12-13.
[75]Ibid., 13-14.

equatorial areas of Mars, they cannot be more than 5 per cent as strong as in the earth's atmosphere and are probably very much less."[76]

As in laboratory studies of hydrogen, high dispersion showed that some emission lines in the Balmer series in stellar spectra were not single, but had two, three, or more components. This composition appeared especially strongly in the spectrum of Mira, in a spectrogram exposed for five and one-half hours, which brought out faint emission lines, and revealed widening and doubling in various lines. However, the explanation accepted for the multiple character of lines in the hydrogen spectrum differed from the explanations accepted in laboratory physics. Adams accepted Paul Merrill's hypothesis that the separation of emission lines into components was only apparent; in fact absorption lines were superposed on broader emission lines, breaking them up and "dividing them into what appear to be individual lines."[77] In contrast to laboratory studies of hydrogen, where great purity in the light source could be achieved, stars were complex in makeup, and many elements had to be accounted for. Mount Wilson staff also observed "doublets," or double lines, in the spectrum of Betelgeuse, and Adams accepted a hypothesis developed for M-type stars by Lyman Spitzer, that the doubling resulted from motion in the stellar atmosphere, which caused Doppler shifting due to radial (line-of-sight) motions.[78]

High-dispersion studies were especially valuable for bringing out very narrow, or sharp, spectral lines, and the faint absorption lines from interstellar matter fell into this category. Interstellar lines could not be identified entirely from elemental spectra; some arose from chemical compounds. Mount Wilson studies confirmed the existence of CH lines, providing "definite proof of the existence of interstellar lines arising from molecular transitions." They also firmed up an identification of interstellar CN. These studies added to Dunham's earlier detection of titanium.[79]

Adams made it clear at the outset, in his summary of new results, that they relied on a Wood grating, together with Schmidt cameras, in the coudé spectrograph of the 100-inch telescope. Wood had received the Draper Medal in astronomy for his work on resonance radiation, the development of absorption screens, and advances with gratings. The citation declared that

> a modern Wood grating with high concentration of light is one of the most effective instruments of research in stellar spectroscopy.

[76]Ibid., 16-17.
[77]Ibid., 17-19.
[78]Ibid., 19-21.
[79]Ibid., 21-23.

It has made possible the analysis of the spectra of the brighter stars on a large scale, has opened up the almost unexplored ultraviolet region of stellar spectra, and has already led to discoveries of interest regarding the constitution of the gases in interstellar space.[80]

The citation recognized Wood as one of a small number of men who could develop methods and instruments, "the widening application of which no one can adequately foresee."[81] In other words, Wood had evolved a technology of discovery.

He was indeed interested in recognition. When titanium lines were detected in interstellar space, Wood wanted to know the advantages conferred by his products, he wrote, "as about the only fun I get out of ruling gratings is in hearing of the successful employment of them by other investigators." Adams assured him that the blazed instrument was "almost entirely responsible" for detection of the ultraviolet lines, and concluded that prism instruments for stellar spectroscopy would soon be eclipsed. Adams said he always tried to make the role of instruments clear in published research on interstellar lines and the far ultraviolet.[82]

But by the late 1930s, Wood felt that his ingenuity was not sufficiently appreciated, and Adams wished to placate him. Adams suggested that Wood come to the observatory and help mount a battery of replica gratings for studies by Fritz Zwicky. The arrangement might, Adams said, "result in something of a revolution in the determination of radial velocities on a wholesale scale." Adams also took a hand in awarding the Draper Medal to Wood, asking Otto Struve of Yerkes Observatory to read the citation at a meeting of the National Academy of Sciences. Even this measure, however, was not enough for a successor of Rowland. Wood complained after the award that his instruments concentrated light more strongly than stated in the citation. Adams reassured Struve that his presentation was not to be faulted: "Wood is a temperamental 'cuss' and I am glad that the award of the Draper Medal went off as well as it did. One never knows just when one may encounter a projecting corner in Wood's disposition."[83]

Wood's perceived slights did not end with the award. When Zwicky discovered a nova while testing a Wood instrument, Wood felt that newspaper accounts should have mentioned him as well as Zwicky. Adams suggested that Wood worry less about publicity and

[80]Draper Medal citation, Box 72, WSA.
[81]Ibid.
[82]Wood to Adams, 5 January 1937; Adams to Wood, 13 January 1937; Adams to Wood, 2 November 1937, WSA.
[83]Adams to Wood, 6 February 1939, 14 October 1940, 18 October 1940; Otto Struve to Adams, 23 May 1941; Adams to Struve, 26 May 1941, WSA.

more about progress in ruling. But the claims continued. At Harvard, Harlow Shapley became aware of Wood's grievances with Mount Wilson, and discussed them with Adams. The issue had now grown to a question of priority regarding the invention of blazed gratings. Adams stated that shaped grooves were not wholly original with Wood. Rowland's assistant, Lewis Jewell, had performed theoretical work on groove form in about 1900, which Wood discounted; Anderson had continued this work. At Mount Wilson, only one Wood grating was in use; others had been made at the observatory. Even so, Adams said, "In my own publications I have tried to lean over backward in mentioning that Wood made the grating with which most of the high dispersion stellar spectra have been photographed." Shapley, aware that Wood had studied groove form as early as 1910, simply hoped that Wood did not plan to publish an "attack."[84]

Zwicky acknowledged Wood's achievement in creating compound or mosaic gratings for Mount Palomar facilities, including a set of fourteen replica gratings. Wrote Zwicky, "It is one of the supreme testimonies to his skill and perseverance that he succeeded in producing three very good 18-inch diameter mosaic gratings." Wood did not complete a grating, as he had hoped, for the 48-inch telescope at Mount Palomar, but Zwicky reported in the 1950s, "Those of us who are now engaged in the attempt to realize the great master's goal are greatly aided by all of the vivid memories of our contacts with him."[85] By the end of World War II, Wood was about 75 years of age, and the golden age of grating production at Rowland's former department was past. Spectroscopy was no longer the favored discipline of physics, as it had been for much of Wood's career. Spectroscopy found greater use in chemistry and astronomy.

At Mount Wilson, there remained a place for those well-versed in the materialities of acquiring observations. John Anderson's work on creating the 200-inch telescope continued. As the large mirror was figured and refigured, he made frequent reports on the condition of the mirror surface. But when it was finally ready to travel to Mount Palomar, Anderson was about 70 years of age, and the trip to Palomar was, for him, unwise.[86] The creations of one generation passed into the hands of another generation, in which a cosmic physics far beyond Rowland's imagining developed, finding not only similarities between the Earth and the heavens, but vast and challenging differences as

---

[84]Wood to Adams, 26 October 1942; Adams to Wood, 2 November 1942; Harlow Shapley to Adams, 26 January 1944; Adams to Shapley, 2 February 1944; Shapley to Adams, 29 April 1944, WSA.

[85]F. Zwicky, *Morphological Astronomy* (Berlin: Springer-Verlag, 1957), 232.

[86]Interview with Ira Bowen, 1968-69, Sources for History of Modern Astrophysics, American Institute of Physics, College Park, MD, 33, 44.

well. Rowland and his followers did not know in advance what their instruments would reveal, but by concentrating on the means of discovery, rather than solely their own expectations, they opened a way of knowing the unknown, and so the implications of their work were more far-reaching.

Bibliography

Abbott, C.G. "Recent Progress in Astrophysics." In *Annual Report of the Board of Regents of the Smithsonian Institution, 1913*, 175-194. Washington: GPO, 1914.

Ames, Joseph S. *Text-Book of General Physics*. New York: American Book Co., 1904.

_____. *The Constitution of Matter*. Boston: Houghton Mifflin, 1913.

Anderson, John A. "The Spectrum of Electrically Exploded Wires." *Astrophysical Journal* 51 (1920): 37-48.

_____. "False Spectra from Diffraction Gratings: Part III. Periodic Errors in Ruling Machines." *Journal of the Optical Society of America and Review of Scientific Instruments* 6 (1922): 434-442.

"Action of the Editorial Board of the Astrophysical Journal with Regard to Standards in Astrophysics and Spectroscopy." *Astrophysical Journal* 3 (1896): 1-3.

Babcock, Horace W. "Diffraction Gratings at the Mount Wilson Observatory." *Vistas in Astronomy* 29 (1986): 153-174.

Bailyn, Bernard. "The Challenge of Modern Historiography," *The American Historical Review* 87 (1982): 1-24.

Ben-David, Joseph. *The Scientist's Role in Society: A Comparative Study* (Chicago: University of Chicago Press, 1984).

Bohr, Niels. "On the Constitution of Atoms and Molecules." *Philosophical Magazine* 26 (1913): 1-25.

Brackett, F.S. "A New Series of Spectrum Lines." *Nature* 109 (1922): 209.

Bridgman, P.W. "Theodore Lyman." *Biographical Memoirs of the National Academy of Sciences* 30 (1957): 237-256.

Campbell, W.W., and Joel Stebbins. "Report on the Organization of the International Astronomical Union." *Proceedings of the National Academy of Sciences* 6 (1920): 349-396.

Catalan, Miguel A. "Series and Other Regularities in the Spectrum of Manganese." *Philosophical Transactions of the Royal Society of London, Series A* 223 (1923): 127-173.

Coben, Stanley. "The Scientific Establishment and the Transmission of Quantum Mechanics to the United States, 1919-32." *American Historical Review* 76 (1971): 442-466.

Crawford, Elisabeth. "The Universe of International Science, 1880-1939." In *Solomon's House Revisited: The Organization and Institutionalization of Science*, ed. Tore Frangsmyr, 251-269. Canton, MA: Science History Publications, 1990.

Crew, Henry. *Laboratory Instructions in Physics: Course A*. Privately printed, 1898.

_____. "Recent Advances in the Teaching of Physics." *Science* 19 (1904): 481-488.

_____. "Fact and Theory in Spectroscopy." *Science* 25 (1907): 1-12.

DeVorkin, David H. "Community and Spectral Classification in Astrophysics: The Acceptance of E.C. Pickering's System in 1910." *Isis* 72 (1981): 29-49.

DeVorkin, David H. and Ralph Kenat. "Quantum Physics and the Stars (I): The Establishment of a Stellar Temperature Scale." *Journal for the History of Astronomy* 14 (1983): 102-132.

_____. "Quantum Physics and the Stars (II): Henry Norris Russell and the Abundances of the Elements in the Atmospheres of the Sun and Stars." *Journal for the History of Astronomy* 14 (1983): 180-221.

Dingle, Herbert. "A Hundred Years of Spectroscopy." *The British Journal for the History of Science* 1 (1963): 199-216.

Foote, Paul D. "William Frederick Meggers." *Biographical Memoirs of the National Academy of Sciences* 41 (1970): 319-340.

Foote, Paul D. and William Meggers. "Atomic Theory and Low Voltage Arcs in Caesium Vapour." *Philosophical Magazine* 40 (1920): 80-97.

Forman, Paul. "The Doublet Riddle and Atomic Physics *circa* 1924." *Isis* 59 (1968): 156-174.

_____. "Alfred Lande and the Anomalous Zeeman Effect, 1919-1921." *Historical Studies in the Physical Sciences* 2 (1970): 153-236.

_____. "Louis Carl Friedrich Paschen." *Dictionary of Scientific Biography*. Charles Coulston Gillispie, ed. Vol. 10 (1974).

Forman, Paul, John L. Heilbron, and Spencer Weart, "Physics *circa* 1900: Personnel, Funding, and Productivity of the Academic Establishments," *Historical Studies in the Physical Sciences* 5 (1975): 1-185.

Forsterling, K., and G. Hansen. "Zeemaneffekt der roten und blauen Wasserstofflinie." *Zeitschrift für Physik* 18 (1923): ** .

Franklin, Allan. *The Neglect of Experiment* (Cambridge: Cambridge University Press, 1986).

_____. "The Epistemology of Experiment," in David Gooding, Trevor Pinch, and Simon Schaffer, eds., *The Uses of Experiment: Studies in the Natural Sciences* (Cambridge: Cambridge University Press, 1989), 437-467.

Frost, Edwin B. *An Astronomer's Life*. New York: Houghton Mifflin Co., 1933.

Gaul, Harriet A., and Ruby Eiseman. *John Alfred Brashear: Scientist and Humanitarian, 1840-1920*. Philadelphia: University of Pennsylvania Press, 1940.

Geison, Gerald L. "Scientific Change, Emerging Specialties, and Research Schools." *History of Science* 19 (1981): 20-40.

Goodstein, Judith R. *Millikan's School: A History of the California Institute of Technology*. New York: W.W. Norton & Co., 1991.

Hale, George Ellery. "The Astrophysical Journal." *Astrophysical Journal* 1 (1895):**.

_____. "The Yerkes Observatory of the University of Chicago. III. The Instrument and Optical Shops, and the Power House." *Astrophysical Journal* 5 (1897): 310-317.

_____. "James Edward Keeler." *Science* 12 (1900): 353-357.

_____. "Opportunities for Solar Research." *Science* 25 (1907): 613-617.

_____. *Ten Years' Work of a Mountain Observatory.* Washington: Carnegie Institution of Washington, 1915.

_____. *National Academies and the Progress of Research.* Lancaster, PA: The New Era Printing Co., 1915.

Hale, George Ellery, and C.D. Perrine. "International Cooperation in Solar Research." *Science* 20 (1904): 930-932.

Hannaway, Owen. "The German Model of Chemical Education in America: Ira Remsen at Johns Hopkins (1876-1913)." *Ambix* 23 (1976): 145-164.

Harwit, Martin. *Cosmic Discovery: The Search, Scope, and Heritage of Astronomy.* New York: Basic Books, 1981.

Hearnshaw, J.B. *The Analysis of Starlight: One Hundred and Fifty Years of Astronomical Spectroscopy.* Cambridge: Cambridge University Press, 1990.

Herzberg, Gerhard. *Atomic Spectra and Atomic Structure.* Translated by J.W.T. Spinks. New York: Dover Publications, 1945. (First published in English 1937.)

Holton, Gerald. "On the Hesitant Rise of Quantum Physics Research in the United States." In *The Michelson Era in American Science 1870-1930*, ed. Stanley Goldberg and Roger H. Stuewer. New York: American Institute of Physics, 1988.

Houston, William V. "The Fine Structure and the Wave-Lengths of the Balmer Series." *Astrophysical Journal* 64 (1926): 81-92.

Hufbauer, Karl. *Exploring the Sun: Solar Science since Galileo.* Baltimore: Johns Hopkins University Press, 1991.

Hund, Friedrich. *The History of Quantum Theory.* New York: Barnes & Noble, 1974.

Kargon, Robert H. *The Rise of Robert Millikan: Portrait of a Life in American Science.* Ithaca: Cornell University Press, 1982.

_____. "Henry Rowland and the Physics Discipline in America." *Vistas in Astronomy* 29 (1986): 131-136.

Keeler, James E. "The Importance of Astrophysical Research and the Relation of Astrophysics to Other Physical Sciences." *Astrophysical Journal* 6 (1897): 271-288.

Kenat, Ralph, and David H. DeVorkin. "Quantum Physics and the Stars (III): Henry Norris Russell and the Search for a Rational Theory of Stellar Spectra." *Journal for the History of Astronomy* 21 (1990): 157-186.

Kent, Norton A., Lucien B. Taylor, and Hazel Pearson. "Doublet Separation and Fine Structure of the Balmer Lines of Hydrogen." *Physical Review* 30 (1927): 266-283.

Kevles, Daniel J. "George Ellery Hale, the First World War, and the Advancement of Science in America." *Isis* 59 (1968): 427-437.

_____. "'Into Hostile Political Camps': The Reorganization of International Science in World War I." *Isis* 62 (1971): 47-60.

_____. *The Physicists: The History of a Scientific Community in Modern America.* Cambridge, MA: Harvard University Press, 1987.

Kragh, Helge. "The Fine Structure of Hydrogen and the Gross Structure of the Physics Community, 1916-26." *Historical Studies in the Physical Sciences* 15 (1985): 67-125.

Kuhn, Thomas S. *Black-Body Theory and the Quantum Discontinuity, 1894-1912.* Oxford: Clarendon Press, 1978.

_____. *The Structure of Scientific Revolutions*, 2d ed. Chicago: University of Chicago Press, 1975. A gathering of anomalies.

_____. *The Essential Tension: Selected Studies in Scientific Tradition and Change.* Chicago: University of Chicago Press, 1977.

Langley, Samuel. [On the distribution of energy in the solar spectrum.] *Nature* 26 (1882): ** .

Lankford, John. "Amateurs and Astrophysics: A Neglected Aspect in the Development of a Scientific Specialty." *Social Studies of Science* 11 (1981): 275-303.

Laporte, Otto, and William Meggers. "Some Rules of Spectral Structure." *Journal of the Optical Society of America and Review of Scientific Instruments* 11 (1925): 459-463.

Latour, Bruno. *Science in Action: How to Follow Scientists and Engineers through Society.* Cambridge, MA: Harvard University Press, 1987.

Livingston, Dorothy Michelson. *The Master of Light: A Biography of Albert A. Michelson.* New York: Charles Scribner's Sons, 1973.

Lyman, Theodore. "False Spectra from the Rowland Concave Grating." *The Physical Review* 12 (1901): 1-13.

_____. "An Explanation of the False Spectra from Diffraction Gratings." *The Physical Review* 5 (1903): 257-266.

_____. "An Extension of the Spectrum in the Extreme Ultra-Violet." *Nature* 93 (1914): 241.

_____. "A Further Extension of the Spectrum." *Nature* 95 (1915): 343.

_____. The Limit of the Spectrum in the Ultra-Violet." *Science* 45 (1917): 187.

_____. "A Helium Series in the Extreme Ultra-Violet." *Science* 50 (1919): 481.

_____. "The Spectrum of Helium in the Extreme Ultraviolet." *Science* 56 (1922): 167-168.

_____. "The Spectroscopy of the Extreme Ultra-Violet." *Journal of the Franklin Institute* 201 (1926): 553-562.

McGucken, William. *Nineteenth-Century Spectroscopy: Development of the Understanding of Spectra, 1802-1897.* Baltimore: The Johns Hopkins University Press, 1969.

Meadows, A.J. *Science and Controversy: A Biography of Sir Norman Lockyer.* Cambridge, MA: MIT Press, 1972.

Meggers, William F. "The Periodic Structural Regularities in Spectra as Related to the Periodic Law of the Chemical Elements." *Proceedings of the National Academy of Sciences* 11 (1925): 43-47.

_____. "Spectroscopy, Past, Present, and Future." *Journal of the Optical Society of America* 36 (1946): 431-448.

Meggers, William F., and C.C. Kiess. "False Spectra from Diffraction Gratings: Part I. Secondary Spectra." *Journal of the Optical Society of America and Review of Scientific Instruments* 6 (1922): 417-429.

Merz, John Theodore. *A History of European Thought in the Nineteenth Century.* 4 vols. New York: Dover Publications, 1975. First published 1904-1912.

214

Michelson, Albert A. "Recent Progress in Spectroscopic Methods." *Nature* 88 (1912): 362-365.

Michelson, Albert A., and Edward W. Morley. "On a Method of Making the Wavelength of Sodium Light the Actual and Practical Standard of Length." *Philosophical Magazine* 24 (1887): 463-466.

Millikan, Robert A. "Radiation and Atomic Structure." *The Physical Review* 10 (1917): 194-205.

Moyer, Albert E. "History of Physics," *Osiris* 1 (1985): 163-182.

Osterbrock, Donald E. *James E. Keeler: Pioneer American Astrophysicist*. Cambridge: Cambridge University Press, 1984.

_____. "Failure and Success: Two Early Experiments with Concave Gratings in Stellar Spectroscopy." *Journal for the History of Astronomy* 17 (1986): 119-129.

Pais, Abraham. *Inward Bound: Of Matter and Forces in the Physical World*. New York: Oxford University Press, 1986.

Parshall, Karen Hunger . "America's First School of Mathematical Research: James Joseph Sylvester at The Johns Hopkins University 1876-1883." *Archive for History of Exact Sciences* 38 (1988): 153-196.

Paschen, Friedrich. "Zur Kenntnis ultraroter Linienspektra." *Annalen der Physik* 27 (1908): ** .

_____. "Bohrs Heliumlinien." *Annalen der Physik* 50 (1916): 901-940.

Paschen, Friedrich and E. Back. "Normale und Anomale Zeemaneffekte." *Annalen der Physik* 39 (1912): 897-932.

Pfund, A.H. "The Emission of Nitrogen and Hydrogen in the Infrared." *Journal of the Optical Society of America and Review of Scientific Instruments* 9 (1924): 193-196.

Pickering, Edward C. in Howard Plotkin, "Edward Charles Pickering's Diary of a Trip to Pasadena to Attend Meeting of Solar Union, August, 1910." *Southern California Quarterly* 60 (1978): 29-44.

Rabkin, Yakov M. "Technological Innovation in Science: The Adoption of Infrared Spectroscopy by Chemists." *Isis* 78 (1987): 31-54.

Reingold, Nathan, ed. *Science in Nineteenth-Century America: A Documentary History*. London: Macmillian and Co., 1966.

_____. *Science, American Style*. New Brunswick: Rutgers University Press, 1992.

Reingold, Nathan, and Ida H. Reingold, eds. *Science in America: A Documentary History, 1900-1939*. Chicago: University of Chicago Press, 1981.

Rothenberg, Marc. "History of Astronomy," *Osiris* 1 (1985): 117-131.

Rowland, Henry A. "Preliminary Table of Solar Spectrum Wave-Lengths. I." *Astrophysical Journal* 1 (1895): ** .

_____. *The Physical Papers of Henry Augustus Rowland*. Baltimore: The Johns Hopkins University Press, 1902.

Rowland. "On the Motion of a Perfect Incompressible Fluid when no Solid Bodies are Present." *American Journal of Mathematics* 3: 226-268.

Rowland. "On the Propagation of an Arbitrary Electro-Magnetic Disturbance, on Spherical Waves of Light and the Dynamical Theory of Diffraction." *American Journal of Mathematics* 6, no. 4: 1-23.

Rowland. *A Preliminary Table of Solar Spectrum Wave-Lengths*. Chicago: The University of Chicago Press, 1898. [Reprinted from several issues of *Astrophysical Journal*.]

Runge, Carl. "False Spectra from Diffraction Gratings: Part II. Theory of Lyman Ghosts." *Journal of the Optical Society of America and Review of Scientific Instruments* 6 (1922): 429-434.

Runge, Carl and F. Paschen. "On the Spectrum of Cleveite Gas." *Astrophysical Journal* 3 (1896): 4-28.

Schaffer, Simon. "Glass Works: Newton's Prisms and the Uses of Experiment," in David Gooding, Trevor Pinch, and Simon Schaffer, eds., *The Uses of Experiment: Studies in the Natural Sciences*. Cambridge: Cambridge University Press, 1989, 67-104.

Russell, H.N., and Frederick A. Saunders. "New Regularities in the Spectra of the Alkaline Earths." *Astrophysical Journal* 61 (1925): 38-69.

Saunders, Frederick A. "Some Aspects of Modern Spectroscopy." *Science* 59 (1924): 47-53.

Schroeder-Gudehus, Brigitte. "Challenge to Transnational Loyalties: International Organizations after the First World War." *Science Studies* 3 (1973): 93-118.

_____. *Les Scientifiques et la Paix: La Communaute Scientifique Internationale au Cours des Annes 20*. Montréal: Les Presses de l'Université de Montréal, 1978.

Schweber, S.S. "The Empiricist Temper Regnant: Theoretical Physics in the United States, 1920-1950." *Historical Studies in the Physical and Biological Sciences* 17 (1986): 55-98.

Seabrook, William. *Doctor Wood: Modern Wizard of the Laboratory*. New York: Harcourt, Brace and Co., 1941.

Servos, John W. "Mathematics and the Physical Sciences in America, 1880-1930." *Isis* 77 (1986): 611-629.

Smith, Robert W. *The Expanding Universe: Astronomy's 'Great Debate' 1900-1932*. Cambridge: Cambridge University Press, 1982.

Sommerfeld, Arnold. *Atomic Structure and Spectral Lines*. Translated by Henry L. Brose. London: Methuen & Co., 1923.

Sopka, Katherine Russell. *Quantum Physics in America: The Years Through 1935*. New York: American Institute of Physics, 1988.

_____, ed. *Physics for a New Century: Papers Presented at the 1904 St. Louis Congress*. New York: Tomash Publishers/American Institute of Physics, 1986.

Stark, J., and H. Kirschbaum. "4. Observations of the Effect of an Electrical Field on Spectral Lines. IV. Types of Lines, Spreading." *Annalen der Physik* 43 (1914).

Stratton, F.J.M. "International Co-operation in Astronomy: A Chapter of Astronomical History." *Monthly Notices of the Royal Astronomical Society* 94 (1934): 361-372.

White, H.E. "Hyperfine Structure in Singly Ionized Praseodymium." *Physical Review* 34 (1929): 1397-1403.

Wood, Robert W. "An Artificial Representation of a Total Solar Eclipse." *Nature* 63 (1901): 250-251.

_____. "The Optical Properties of Metallic Vapors." In *Lectures Delivered at the Celebration of the Twentieth Anniversary of the Founding of Clark University*, 96-127. Worcester, MA: Clark University, 1912.

216

_____. "Resonance Spectra of Iodine by Multiplex Excitation." *Philosophical Magazine* 24 (1912): 673-693.

_____. *Physical Optics.* New York: The Macmillan Co., 1911 (second edition); reprinted 1924.

_____. "Hydrogen Spectra from Long Vacuum Tubes." *Philosophical Magazine* 42 (1921): 729-745.

_____."Atomic Hydrogen and the Balmer Series Spectrum." *Philosophical Magazine* 44 (1922): 538-546.

_____. "An Experimental Study of Grating Errors and 'Ghosts.'" *Philosophical Magazine* 48 (1924): 497-508.

Wood and R. Fortrat. "The Principal Series of Sodium." *Astrophysical Journal* 43 (1916): 73-80.

Wright, Helen. *Explorer of the Universe: A Biography of George Ellery Hale.* New York: E.P. Dutton & Co., 1966.

Wright, Helen, Joan N. Warnow, and Charles Weiner, eds. *The Legacy of George Ellery Hale.* Cambridge, MA: The MIT Press, 1972.

Zeeman, P. "On the Influence of Magnetism on the Nature of the Light Emitted by a Substance." *Philosophical Magazine* 43 (1897): ** .

Archival Sources

Archive for History of Quantum Physics.

Walter S. Adams Papers. Carnegie Observatories Collection, Huntington Library, San Marino, California.

Henry Crew Papers. Archives, Northwestern University, Evanston, Illinois.

Henry Crew Papers, Microfilm. Niels Bohr Library, American Institute of Physics, New York, New York.

Daniel Coit Gilman Papers. Ms., Special Collections, Milton Eisenhower Library, Johns Hopkins University, Baltimore, Maryland.

George Ellery Hale Collection. Huntington Library, San Marino, California.

George Ellery Hale Papers. Carnegie Observatories Collection, Huntington Library, San Marino, California.

George Ellery Hale Papers, Microfilm.

Theodore Lyman Papers. Archives, Pusey Library, Harvard University, Cambridge, Massachusetts.

William F. Meggers Papers. Niels Bohr Library, American Institute of Physics, New York, New York.

Paul Willard Merrill Papers. Carnegie Observatories Collection, Huntington Library, San Marino, California.

Henry A. Rowland Papers. Ms. 6, Special Collections, Milton Eisenhower Library, Johns Hopkins University, Baltimore, Maryland.

Robert W. Wood Papers. Ms. 96, Special Collections, Milton Eisenhower Library, Johns Hopkins University, Baltimore, Maryland.

Robert W. Wood Papers. Niels Bohr Library, American Institute of Physics, New York, New York.

Index

Entries in italics represent figures; "t" following an entry represents a table; "n" denotes footnote; HAR = Henry A. Rowland.

A

Abetti, Giorgio, 153
Abney, William, Sir, 48
absolute measurements, 118–119
acoustical analogy
 with atomic physics, 168
 with diffraction gratings, 30–31
 with light-emitting objects, 157
 with light spectra, 183–184
Adams, Joseph, 192
Adams, Walter, 192, 203–204
 revision of HAR's solar wavelength
  tables by, 196–198
 spectroscopic innovations at Mount
  Wilson Observatory and,
  205–208
Allegheny Observatory (PA), 43–44, 94,
  102, 103
Alpha Orionis, 199
American Academy of Arts and Sciences, 57
American Association for the Advancement of Science, 25
American Journal of Mathematics, 116
American Journal of Science, 6, 79
American Physical Society, xvii,
  xviii–xix, xxiii

Ames, Joseph S., xviii, xx, 25, 58,
  107–108, 109, 116, 138
 acoustical analogy with light-emitting objects of, 157
 Anton Grünwald's critique of, 61–62
 career course of, 63–64
 conservatism of, 155–156
 editorial pursuits of, 94, 95
 Germanic influences exhibited by,
  154
 gradualist view of, 153
 hydrogen studies of, 63–64
 laboratory spectral studies of, 74
 publications of, 153
 theory of relativity and, 155–156
 vs Robert Wood, 155
Anderson, John A., 100, 143, 188
 employment at Mount Wilson Solar
  Observatory of, 108–112, 111,
  132
  accomplishments during,
   193–194, 198–200
 ghosts and, 134
 grating production by, 132
 laboratory studies of, 149–152
  grating production for, 151
  high voltage, 150–151, 193–194,
   198
  Stark effect, 149–150

mechanical skills of, 23–24
Mount Palomar 200-inch reflecting
telescope work of, 151–152,
199–200, 209
Angström, Anders, xxiii, 20–21, 34
Anthony, William A., xix
"A Plea for Pure Science," xxiii, 24
Arcturus, 105, 108, 148, 154, 200, 201, 206
Astronomische Gesellschaft, 161
Astronomische Nachrichten, 160
astronomy, 176
ghosts and, 117
laboratory spectroscopy and,
188–189
physics and, 189
union of (*See* astrophysics)
Astronomy and Astrophysics, 94–95
astronomy of the atom, 188
Astrophysical Journal, 22, 85, 95–98,
159–160, 160
George Ellery Hale as founder of,
78–79
George Ellery Hale's contributions
to, 96–98, 96n
publication avenue of School of
Light and, 78–79
astrophysics, 93, 94, 95, 111–112
George Ellery Hale's views on,
108–109
James Keeler's views on, 101, 102,
106–107
rise of, 99–102
solar physics and, 99, 100
Atomic Structure and Spectral Lines
(Sommerfeld), 175
atomic theory
acoustical analogy with, 168
HAR's views on, 114–116
Niels Bohr's views on, 125
William Meggers' spectroscopic
studies and, 130–131, *131*, 175
atomo-mechanism, xxi
aurora borealis (northern lights), 3, 11,
12

B

Babcock, Harold D., 203
Back, Ernst, 171–172
Bacon, Francis, 185

Baird, Walter, 147
Baird Associates, early commercial
spectrographs of, 146–147
Balmer formula, 63, 120, 124
Balmer series, 128–129, 207
hydrogen spectrum of, 169–172
sodium spectrum of, 127–128
Barker, George, 11
Barnes, James, 73
Bausch & Lomb, 147, 202
Bazerman, Charles, 141–142
Bedell, Frederick, xix
Bell, Alexander Graham, 25, 83n
Bethe, Hans, 180
Big Spectroscope, 22n, 54–58, 61–62, 68,
74, 89. *See also* spectroscope
Birge, Raymond, 130
blackbody radiation, 19, 19n, 46, 73
Bohr, Niels, 125
Bohr atom, 114–115, 168
Bohr's theory, 130, 179–180, 184
bolometer, 44, 102
Bond, George M., xxiv
Born, Max, 180
Boscovich, R.J., 115
Boston University, 173
Bowen, Ira, 132, 152, 188–189, 198, 202
image-slicer of, 206
nebular studies of, 105
Brace, DeWitt, 70
Brackett, F.S., 169
Brashear, John, xxiv, 48–50, 50, 91, 103
instrument construction by, 41–43
James Keeler and, 102–103
official Rowland grating marketing
agent, 49
Bruce Fund, 57
Bunsen, Robert, 19–20, 26
Bureau of Standards, 158–159, 202
staff of, 155
William Meggers' work at, 177–180

C

C.A. Steinheil Sohne, 42n
California Institute of Technology,
144–145, 173
fine structure studies at, 188
Cambridge atom, 105
Cambridge University, xxiv, 105, 168

Campbell, W.W., 97
Capella, 199
Carnegie, Andrew, 97
Carnegie Institution of Washington, 97, 98
Catalan, Miguel A., 170, 171
Cavendish Laboratory, xxiv, 105
Celestial-terrestrial studies
    of Henry Crew, 148, 149
    of School of Light, 22, 148–152
cesium, 130, *131*
charisma of HAR, 69–71
chemical replication, 143
Chicago Columbian Exposition, 104–105
Clark, Alvan, 101
Clark University, xviii
College of Wooster, 4
collimator for Mount Wilson 100-inch telescope, 202
Columbia University, xvii, xx
condensing lens, 12n
conservatism, 25
    of Johns Hopkins University physics department School of Light, 138–139, 152–157
    of Joseph Ames, 155–156
Cornell University, xix
Corning, New York, 151
Cornu, M.A., 95, 159–160
coudé focus of 100-inch Mount Wilson telescope, 203
Crawford, Elisabeth, 139–140, 141, 159
Crew, Henry, 26–27, 57, 64, 74, 157, 173
    celestial-terrestrial studies of, 148, 149
    classical physics memoirs of, 153–154
    comment:, 68, 157–158
    editorial pursuits of, 94, 95, 153–154
    employment at Northwestern University, 148
    HAR and, 70–71
    James Keeler and, 103
    Saturnian atom and, 168–169
    wave theory of light and, 153
crystals, as x-ray diffraction instruments, 174

D

Daston, Lorraine, 139
de Candolle, Alphonse, 2n
de Salvio, Alfonso, 153
Deslandres, Henri, 92, 160
Dialogues Concerning Two New Sciences (Galileo), 153
diamond point, grating production and, 38
discharge tube, 123
dissociation hypothesis, 52
doctrine of conservation of force (energy), 2–3
Dow Chemical Company, 147
Draper, Henry, 83n
Draper, John William, 13–14
Draper, John William (father), 83n
Draper Medal, 207–208
drive belt, 135
    ghost formation and, 134
Duner, N.C., 95
Dunham, Theodore, 203, 205

E

earth, 57, 197
echelette instruments, 143–144
echelon instrument, 144
Eder, J.M., 158
Edison, Thomas Alva, 11, 24–25, 40
Einstein, Albert, 194–195, 194n
Einstein collaborators, 195
electrodes, spectral lines and, 123
Eliot, Charles, 14–15
engineering
    electrical and mechanical, prominence in physics of, xx–xxi
    HAR's training in, 3–4, 137
    influence on American physicists, xvi–xvii
    laboratory, xxii, xxiv
Enlightenment, 160
Epstein, Paul, 180

F

Faraday, Michael, 6, 115
Fehrenbach, Charles, 204–205
fog
    in laboratory of Theodore Lyman,
        122
    on photographic plates, 123
Foote, Paul, 130
forbidden behavior of electrons, 205
forbidden lines, 202–203
Forman, Paul, 175
Fraunhofer lines, 96–97, 138. *See also*
        von Fraunhofer, Joseph
Fresnel, Augustin, 13

G

Galileo, 153
Gamow, George, 180
German V-2 rocket, Rowland grating
        mounted in, 145–146
Germany, xxiv, 8, 16, 148, 173, 175–177
    HAR's travels in, 7–8
    influence on Joseph Ames, 154
    James Keeler's studies in, 102, 103
    U.S. graduate students and, 22–23
ghosts, grating, 131, 135–136
    Carl Runge's explanation of, 133–134
    cause of, 114
    defined, 113–114
    drive belt role in, 134
    HAR's background knowledge of,
        116–117
    John Anderson's work with, 134
    origin of term, 117
    phantoms vs, 120
    Rowland, 132, 133
    Theodore Lyman and, 132, 133
Gibbs, Josiah Willard, xvi–xvii, 5
Gilman, Daniel Coit, xxii, 5, 47, 49, 50,
        81
    comment: academic freedom in
        Johns Hopkins University
        physics department, 72
    comment: biology/physiology in
        medical training, 82
    comment: HAR's ministerial her-
        itage, 67

hiring of HAR by, 7
James Keeler's contact with, 104
graduate training at Johns Hopkins
        University, 14
grand piano analogy to atomic
        physics, 168
grating(s), *39*
    blazed, 203–205, 209
    concave diffraction (*See also* grat-
        ing(s), Rowland)
        analogy with sound of, 30–31
        Jacomini, 201–202, 203
        Johns Hopkins University
            physics department, cap-
            tured German V-2 rocket
            carrying, 145–146
        Joseph von Fraunhofer's pro-
            duction of, 31–34
        Kenwood Physical Observa-
            tory and, 91
        map creation with, 52–53
        marketing of, 48–50
        operation of, 38–40
        overlapping spectra produced
            by, 40–41
        Pasedena-produced (Caltech),
            132
        power for, 38
        prisms vs, 12–13, 13, 30, 31
            difficulties associated with,
                13
        reflection, 32
        ruling engines to produce,
            34–36
        stellar spectroscopy and,
            201–209
        Theodore Lymans' critique of,
            114
        use by Ernst Back, 171–172
        use by Friedrich Paschen,
            171–172
        vs interferometer, 167
    flat, 14
    glass, 144
    Michelson, 123–124
    prisms vs, 147
    Rowland, 16, 22, *55* (*See also* grat-
        ing(s), concave diffraction)
        construction of, 13–14, 36–41
        line densities in, 43
        near perfection in, 46–47

precision surface formation in, 41–43
Theodore Schneider's role in, 49–50
contribution to physics of, 146–147, 170, 171
contribution to spectroscopy, 20, 43–44
critique of, 51–52
dissemination of, 48–51, 107
echelon, 31
first, 31
ghosts in, 119
Johns Hopkins University physics department's spectroscope and, 54, 56–57
U.S. sources of, 132
Rutherfurd, 14
stellar spectroscopy and, 148–149
transmission, 32
U.S. sources of, 134–135
Grünwald, Anton, 61–62

H

Hale, George Ellery, 78–79, 85–98, 148
Astrophysical Journal contributions of, 96–98, 96n
astrophysics and, 108–109
crossover from laboratory to observatory plans of, 189
education of, 88, 100, 101
HAR's meeting of, 89
HAR's solar wavelength table revisions of, 189–192
home spectroscopic laboratory of, 89–90
international ties of, 92, 190
laboratory reproductions of celestial conditions by, 149
observatories established by, 90–94
(*See also* specific observatories)
impact of Hopkins gratings on, 107
publications of, 94–98
Rowland grating and, 87–88
Smithsonian Institution position and, 98, 98n
solar magnetism studies of, 107, 107n

solar spectroscopy career of, 88–89
spectrohelograph invented by, 90
stellar spectra interests of, 200–201
telescope built by, 148–149
at University of Chicago, 92
Hale, Mary, 88
Hale, William Ellery, 88
Hall, Edwin, xviii, xxii, 9, 125
Hall Effect, xxii
Hall of Science for the Century of Progress Exposition (Chicago, 1933), 154
Hannaway, Owen, comment: Ira Remsen, 81
Harper, William, 92
Harvard Astronomical Observatory, 57, 100
Harvard University, 94
Harwit, Martin, comment: rise of astrophysics, 99
Hastings, Charles, xviii, 41, 68, 83n, 91, 95
Heisenberg, Werner, XV, xv–xvi, xx–xi, 179
heliostat, 53
helium, 126–127
Hentschel, Klaus, 21n, 138
Hoag, Baron, xv, xx–xi
Hopkins, John, university endowment of, 82
Hounshell, David, 24
Houston, William, 173
Howell, Janet, 110, 110n
Huggins, William, 92–93, 95, 96, 105, 160
humility, 65–66, 66
Humphreys, William, 73, 74, 157
Hund, Friedrich, 179
hydrogen, 123
Joseph Ames' studies of, 63–64
solar quantities of, 189, 195
spectrum(a) of, 114, 169–172
fine structure of, 170–173, 175
hyperfine structure of, 173
Robert Wood's studies of, 128–129
solar quantities of, 189, 195
Theodore Lyman's series of, 119–120

I

image-slicer, 206
informality of Johns Hopkins University physics department, 72
infrared spectrum, 44, 102, 169–170
instruments. *See also* specific instruments
    HAR's Rensselaer collection of, 3–4
    importance to HAR of, 3–4, 9–11, 43–44, 100
    Michelson's views on, 167–168
    Rowland
        Johns Hopkins University School of Light and, 22, 52–53, 83, 142–147
        physics applications of, 146–147, 170–171
interferometer, 167, 172n, 199
International Astronomical Union, 159, 161, 176
International Electrical Congress, xix–xx
internationalism, 6–8, 139–141, 157–162
    of American physicists, xvii–xviii
    of Charles St. John, 190
    of HAR, xxi–xxii
    HAR's students and, 22–23, 157–158
    of William Meggers, 175–176
International Research Council, 111, 161
International Union for Cooperation in Solar Research, 118n, 160

J

Jacomini, Clement, 110–111
Jacomini grating, 201–202, 203
Janssen, Jules, 50
Jefferson Laboratory, 126
Jewell, Lewis, 20–21, 52–53, 57, 73, 74–75, 143
    grating production for George Ellery Hale of, 107–108
    priority of blazed gratings and, 209
Johns Hopkins University. *See also* research schools
    endowment for, 82
    graduate training at, 14

organizational development of, 5
physics department of
    diffraction instruments produced at, 22, 52–53 (*See also* grating(s), Rowland)
    HAR as head of
        academic freedom and, 72
        financial support for, 83–84
        influence on rise of astrophysics with, 100
        informality and, 72
        institutional power of, 72
        light experiments with, 11–12
        ministerial heritage of, 67–68
        noteworthy studies of, 9–11
        recruitment by Daniel Coit Gilman of, 7
        research program of, 74–76
        Science Association and, 74
        students under, 72–75
    Henry Crew as philosopher of, 26–27
    history of, 15–16
    influence on spectrum analysis of, 120–121
    instruments owned by, 83
    James Keeler's reliance on, 103–104
    Joseph Ames and, 64
    legacy of, 23–24, 25–26
    physicists as product of, 16, 17, 22
    preeminence of, xvi–xvii
    quality of, 17
    research foundations of, 15–16
    School of Light (*See* School of Light)
    study of light at (*See* light; spectrum(a), light)
secular character of, 81
U.S. research schools development and, 81–83
U.S. university development and, 81
Journal of the Optical Society of America, 176

K

Kapteyn, J.C., 100
Kayser, Heinrich, 103, 158

encyclopedias of spectroscopy, 177
Keeler, James, 48, 70, 100–107
    astronomical spectroscopic mile-
        stones of others listed by, 106
    astrophysics and, 101, 106
    editorial pursuits of, 79, 94, 95
    education of, 101–102
    employment of, 102, 103
    European travels of, 102–103
    John Brashear and, 102–103
    nebular studies of, 103–106
    Saturn studies of, 105–106
    stellar studies of, 103
Kemble, Edwin, Robert Wood and, 181
Kent, Norton, 173
Kenwood Physical Observatory, 90–91
Kevles, Daniel, 27
    comment: quality of School of Light
        physics department, 155
    comment: *The Physical Review*, 97
    The Physicists, xx–xxi
Kiess, Carl, 177
King, Arthur, 193
    Henry Norris Russell and, 196
Kirchhoff, Gustav, 19–20, 26
Kohlrausch, F.W.G., 48
Kragh, Helge, 172
Kuhn, Thomas, 46

L

Laboratory of Astrophysics and Physi-
    cal Meteorology, 146
Langley, Samuel, 19n, 42–43, 52, 83n,
    91, 94, 102
    early use of Rowland concave dif-
        fraction grating by, 43–46
Lankford, John, comment: rise of astro-
    physics, 99
Laporte, Otto, 154
    comment: contrasting styles of
        Joseph Ames and Robert
        Wood, 155
lasers, 143
Latour, Bruno, 121n, 122n
Lewis, G.N., 184
Lick Observatory (CA), 94, 97, 148, 192
    James Keeler at, 103
    telescope at, 92
light. *See also* spectrum(a), light

analysis of
    implications of, 12
    spectral lines in, 17–19
artificial sources of, 18
celestial (*See* spectrum(a), light,
    celestial)
diffuse, spectroscopic impurities of,
    123–124
electromagnetic derivation of veloc-
    ity of, 12
historiography of, 19
Johns Hopkins University physics
    department study of, 11–12,
    26, 53–58, 83
terrestrial (*See* spectrum(a), terres-
    trial)
wave theories of, 153
light bulb, electric, 11, 12
lithium chloride, 77–78
Lockyer, J. Norman, 46, 48, 91
    astrophysical work of, 99
    critique of Rowland gratings by,
        51–52
Lorentz, H.A., 180
Lyman, Theodore, 119–127, 135–136,
    144, 173, 188, 196
    critique of Rowland gratings by, 114,
        119
    director of Jefferson Laboratory, 126
    European travels of, 180
    ghosts of, 132
        phantoms vs, 120
    hydrogen spectral series of, 119–120
    laboratory setup of, 122–124
        leaks, impurities, and "fog" in
        the, 122–123
    mathematical approach to false
        lines, 121
    military service of, 126
    nonmathematical approach to
        physics, 119–121
    personal travels of, 125n
    phenomenological approach of,
        124–125
    post-WWI activities of, 126–127
    studies of, formulas and theory and,
        126
    ultraviolet spectra studies of, 125,
        131–132
        publications on, 125
Lyman alpha, 119

M

Machinery Hall, xix
magnesium spectrum, 120
magnetism, 6, 7, 11
 terrestrial, 3, 11n
maps, solar. *See* solar maps
Mars, 200
 analysis of water-vapor lines of,
 206–207
Martin, Henry Newell, 15n, 82
Mascart, E.E., 48
Massachusetts Institute of Technology,
 10, 78
mathematics
 HAR's use of, 5–6, 116
 laboratory instruction in, 82–83
 spectral lines and, 121
Maxwell, James Clerk, xxiii–xxiv, 5, 139
 electromagnetic theory of light, 25,
 139
 recognition of HAR by, 6–7
 Saturn ring studies of, 105–106
measuring engines. *See* ruling engines
Meggers, William, 103, 154, 173, 184
 complex structure studies of, 175
 conservation of Johns Hopkins University physics department
 heritage by, 192
 ghosts and, 133–134
 Henry Norris Russell and, 181–185
 international ties of, 158–159,
 175–176
 European, 176–177
 organizing principles of, 176
 solar mapping activities of, 192–193
 spectra cataloguing by, 131
 spectroscopic work of, 129–131
 at Bureau of Standards,
 177–180
 implications for atomic models
 of, 130
 U.S. contacts of, 176
 use of "resonance" vs Robert Wood,
 179
 wavelength studies of, 194–195
Mendenhall, Charles E., 19n, 98, 110n
 Frederick A. Saunders and, 73
Merrill, Paul, 180
 high-dispersion studies of, 201–203

hypothesis of spectral emission
 lines, 207
Mt. Wilson studies of, 194
solar mapping interests of, 193
Merritt, Ernest, xviii
Merz, J.T., comment: internationalism
 in Western scientific community, 140–141
Meudon Observatory, 50
Michelson, Albert A., xviii, 31, 42, 75,
 95, 188
 engineering background of, xvi–xvii
 spectral line studies of, 166–168
 views on progress of instrumenta-
 tion *vs.* HAR, 167–168
Miller, John, 68
Millikan, Robert, 126, 131, 188–189,
 193, 198
 comment: Theodore Lyman and
 atomic theory, 125
 move to California Institute of Technology of, 132
Mira, 207
Mohler, John, 73, 74–75
Moore, Charlotte, 195, 196–198
Moore, Joseph, 192
Morrell, J.B., 69
Mount Palomar Observatory (CA), 90.
 See also California Institute of
 Technology
 200-inch reflecting telescope at, John
 Anderson's work on, 151–152,
 199–200, 209
Mount Wilson Solar Observatory (CA),
 90, 97–98, 132, 149, 176, 180,
 188
 George Ellery Hale's desire for
 School of Light expertise at,
 107
 grating problems at, 108
 Henry Norris Russell's access to,
 195–196
 100-inch telescope at
 development by Paul Merrill
 of, 201–203
 high-dispersion investigations
 performed on, 206–208
 Robert Wood's grating for,
 203–205
 Theodore Durham's work on,
 205

International Union for Cooperation
  in Solar Research meeting at,
  160–161
John Anderson and, 108–112, 198–200
research done at, 193–196
revision of HAR's solar wavelength
  tables at, 189–192, 196–197,
  196–198
solar data from, 195
spectrographs used at, 198–199
spectroscopic innovations at, 205–208
World War I and, 111
Moyer, Albert, xxi
  American Physics in Transition, xx–xxi
multiplets, 170, 189

N

National Academy of Sciences, 208
National Bureau of Standards, 111, 129,
  154, 181, 193
  School of Light employees of, 80–81
National Research Council, 111
Nature, 46, 99, 124
Nautical Almanac in Washington,
  D.C., 102
Naval Research Laboratory, 145
nebula
  Ira Bowen's studies of, 105
  James Keeler's studies of, 103–106
nebulium, 202
  lines of, 188–189
Newcomb, Simon, 83n, 98n, 102, 116
Newton, Isaac, 18, 113n
Nichols, E.L., xviii
Nobert, Friedrich, 48
  diffraction gratings created by, 34
northern lights (aurora borealis), 3, 11,
  12
Northwestern University, 148

O

Observatory of Solar Physics, 99
Ohio State University, 173
ohm, 9, 71
Osterbrock, Donald, comment: concave gratings and stellar
  spectroscopy, 148–149

P

parallax, solar, 206
Parshall, Karen Hunger, comment:
  James Joseph Sylvester, 82
Pasadena, 132, 135
Paschen, Friedrich, 173, 178
  Arnold Sommerfeld and, 174
  concave grating use by, 171–172
  experimental studies of, 170–171
  neon spectrum studies of, 184
Pauli, Wolfgang, 179
Payne, Cecilia, 195
Payne, W.W., 95
Pease, Francis, 109–110, 199
Peirce, Charles S., 116–119
Perry, Wilbur H., 143, 145, 147
Pfund, August, 145, 161n, 169–170
phantom, ghosts vs, 120
Philosophical Magazine, 6, 79, 158
photographic plates, 54, 122
  emulsions on, 201
  fog on, 123
photography, spectral, 46, 104–105
*Physical Review: A Journal of Experimen-
  tal and Theoretical Physics*,
  xviii, xix
Physical Society of London, 46
physicist(s), product of Johns Hopkins
  University Department of
  Physics, 16
physics
  American vs European, xv–xvi,
    xxiv–xxv, 5–6, 180–185
  cultural values of, xxiii–xxv
  technology of, xxiii–xxv
  astronomy and, 176, 189 (*See also* as-
    trophysics)
  background of, xvii
  classical memoirs of, 153–154
  engineering approach to, xv–xvii
  Galileo and, 153
  HAR's style of, 1, 2, 137
    internationalism and, 6–8,
      22–23, 139–141, 157–162
  history of, 17
  physical discontinuities and, 19
  pre-atomic, 19
  prominence of electrical and mechan-
    ical engineering in, xx–xxi

quantum theory of, 19
solar, 99, 100
spectroscopy and, 168
theoretical vs experimental, 180–185
theory of relativity and, 19
physics department at Johns Hopkins
University. *See* Johns Hopkins
University, physics department of
Physikalisch-Technische Reichsanstalt
(Berlin), 173
Pickering, Edward, 92, 95, 161
Plaskett, J.S., 201
Pratt and Whitney, xxiv
Princeton University, 176, 181
prism(s), 30, 120
glass, 171
gratings vs, 12–13, 30, 31, 147
quartz, 171
Proceedings of the Royal Society, 158
Pulkowa Observatory, 48
Pupin, Michael, xvii, xx
purity, scientific
HAR and, xxiii, xxv, 24–25, 27, 53,
66, 67
Theodore Lyman and, 119–120, 121,
123

Q

quantum physics, 19, 181
quantum theory of spectral line emission, 178
Quincke, Georg, 48

R

raie ultime, 179
Ramsay, William, 75
Rayton, W.B., 202
redshift, gravitational, 192, 192n
Reid, Harry, 73
religion and science, 2
Remsen, Ira, 12, 67, 81–82, 109
Rensselaer Polytechnic Institute
curriculum of (1854), 35–35
HAR at, 4, 66, 119
HAR's instruments at, 3–4
instruction methods at, 4n

research schools. *See also* Johns Hopkins University
laboratory based, 68–85, 69t
resonance line, 179
Rittenhouse, David, 31
Ritz, Walter, 124
Rockwell, Charles, 101
Rogers, William A., xxiv, 14, 42, 116,
119
ruling engines of, 35
Rogers-Bond comparator, xxiv
Rosenberg, Robert, 17
Rossetti, F., 48
Rowland, Henry A., *ii*, 95
acoustical analogy with light-emitting objects of, 157
atomic theory and, 114–116
celestial-terrestrial interests of,
149–152
charisma and, 69–71
classroom inadequacy of, 67–71
education of, 3
employment history, 4
engineering background of, xvi–xvii,
3–4, 137
European travels of, 7–8, 9, 46–48
experiments of, xxii–xxiii
fortunes of, 47
George Ellery Hale and, 89
George Sweetnam's views of,
xxii–xxv
independent research of, 3
instruments of (*See* instruments)
Joseph Ames and, 64–65
mathematics use by, 5–6, 116
ministerial heritage of, 66–67
physics profession in time of, 1–2
physics style of, 1, 2, 137
internationalism and, 6–8,
139–141, 157–162
products of, 16
professional value(s) of
celestial-terrestrial comparisons as, 22
conservatism as, 138–139
engineering approach to
physics as, 137, 137n
hesitancy as, 65
humility and, 65–66, 67
internationalism as, 6–8, 22–23,
139–141, 157–162

laboratory ethos of, 64–65
mechanical facility as, 22–24
scientific conservatism as, 25
scientific purity as, xxiii, xxv,
  24–25, 27, 53, 66, 67
recognition by James Clerk Maxwell,
  6–7
Rensselaer period of (*see* Rensselaer
  Polytechnic Institute)
research success of, 68–71
self-confidence of, xxi
solar wavelength tables of, 20, 21.96,
  118–119, 189–192, 196–197
standard spectrum reference line of,
  118–119
teaching reputation *vs.* research success of, 68–71
terrestrial magnetism studies of, 3,
  11, 11n
views on progress of instrumentation *vs.* Michelson, 167–168
views on proper combination of
  teaching and research, 9
William Rogers and, 35
Rowland, Henry A. (father), 1, 2
Rowland, Henry A. (grandfather), 1, 2
Royal Institution, London, 9
ruling engines, xxiv, 34–36, *39*, 57, 108,
  112, 132, 135
  imperfect functioning of (*See* ghosts,
    grating)
  at Mount Wilson Solar Observatory,
    107–108, 109–110
  at School of Light, 142–143
  thermal and mechanical stability of,
    40
Rumford Fund, 57
Rumford Medal, 47
Runge, Carl, 121, 132, 133–134, 171–172
Russell, Henry Norris, 176
  A.S. King and, 196
  calibration of Rowland intensities
    by, 189–190
  Frederick Saunders and, 181–185
    notation system developed
      through, 182–183
  Mount Wilson Observatory observa-
    tions and, 195–196
  William Meggers and, 181
Rutherfurd, Lewis, 14, 34–35, 116
Rydberg formula, 63, 120

S

Saturn, 105–106
Saturnian atom, 105, 168–169
Saunders, Frederick, 19n, 115n, 135
  Charles E. Mendenhall and, 73
  Henry Norris Russell and, 181–185
  reservations about Bohr model of,
    184
  spectra in analogy with music and,
    183–184
Schmidt camera, 203–204, 205–206
Schneider, Theodore, xxiv, 14, 37, 40
  importance to Rowland grating con-
    struction of, 49–50
School of Light, 94
  celestial-terrestrial studies of, 148–152
  conservatism of, 138–139, 152–157
  elements of style of, 135–141, 137n
  George Ellery Hale and, 89
  instrument-oriented approach of,
    142–147
  internationalism of, 157–162
  inventiveness at, 141–142
  mechanical models of light-emitting
    objects pursued by, 157
  physicists produced by, 146
  research field invaded by, 76
  spectral lines and, 169–170
  spectroscopists produced by, 79–81
  student recruits of, 77–78
    internationalism of, 157–158
  successful laboratory-based research
    school characteristics extant
    at, 68–85, 69t
  university setting of, 81
  vehicles for students' publications
    from, 78–79
Schrödinger, Erwin, 180
Schumann, Victor, 50–51
Science, 25
science and religion, 2, 81–82
Science Association of Johns Hopkins
  University physics department, 74
science historiography, successful laboratory-based research school
  characteristics and, 68–85, 69t
scientific method, Henry Crew's definition of, 26–27

screw, feed, 43
  fine, 54, 56, 57
  grating production and, 31, 36–37,
    145
seismometer, 198–199
Shapley, Harlow, 209
Sidereal Messenger, 95
Sirius B star, 192
Smithsonian Astrophysical Observa-
    tory, 91, 94, 94n
Smyth, Piazzi, 50
sodium, Robert Wood's studies of,
    127–128
solar maps, 16, 21, 26, 52, 57, 61
  Angström's, 76
  Rowland's, 76–77, 118–119
  section from, *45*
solar parallax, 206
solar spectrum. *See* spectrum(a), solar
solar wavelength tables of HAR, 20, 21,
    96, 118–119, 196–197
  revision of, 189–192, 196–197
Sommerfeld, Arnold, 154, 172, 175
  Atomic Structure and Spectral Lines by,
    175
  Friedrich Paschen and, 174
  spectra in analogy with music and,
    183–184
  theoretical studies of, 173–175
sound analogy. *See* acoustical analogy
spectrogram(s), J.M. Eder's collection
    of, 158–159
spectrograph(s)
  direct reading, 147
  early commercial, 146–147
  at Mount Wilson, 205–206, 1980199
spectrohelograph, 90
spectrometer(s), rock-salt, A.H.
    Pfund's studies with, 169–170
spectroscope(s), *56. See also* Big Spec-
    troscope; spectroscopy
  advantages of Mount Wilson inno-
    vations in, 205–208
  obstacles to purity of, 123–124
  problems with in Theodore Lyman's
    laboratory, 122–124
spectroscopists, School of Light, 79–81
spectroscopy, 13, 19–20. *See also* spec-
    trum(a), light
  articles in *The Physical Review* on,
    141–142

celestial studies in, 19–20, 149
conservatism of practitioners of, 152
formulas and theory and, Theodore
    Lyman's work and, 126, 127
Henry Crew's work in, 148, 149
historical periods of, 204–205
history of, 16
hot-spark vacuum and, 188
impact on physics, 168
laboratory, astronomy and, 188–189
ongoing impact of, 165–166
origins of, 31–34
PhDs in, 80t
solar, George Ellery Hale's career in,
    88–89
stellar, 200–209
  concave gratings and, 148–149
  gratings and, 22, 201–209
William Meggers' work at Bureau of
    Standards in, 177–180
spectrum(a), light
  of air, 122–123
  analysis of, 17–19
    automation of, 147
    influence of Johns Hopkins
      University physics depart-
      ment on, 120–121
  blackbody, 19, 19n, 46, 73
  celestial, 32–33 (*See also* entries
    below under solar; stellar)
    laboratory studies of, 149
      high voltage, 150–151,
        193–194, 198
    School of Light, 148–152
  solar, 16, 18, 57–58
    HAR's study of, 20, 21, 74
    John Mohler's studies of, 5
    measurement of, 44–46, 88
    terrestrial *vs.*, 102–103
  stellar, 103–104
  student research on, 75–76
cesium, 130, *131*
continuous, 18
elemental cataloguing of, 131
emission lines, Paul Merrill's hy-
    pothesis of, 207
fine structure studies of, 188
ghosts in (*See* ghosts, grating)
helium, 126–127
infrared region of, 44, 102, 169–170
iodine, 165–166

iron, 151, 202–203
lense absorption of, 203–204
lines in, 17–19
    ghosts (*See also* ghosts, grating)
    presence of chemical element
        and, 19–20
    proliferation of, 165–168
    significance of, 19, 96–97,
        102–103
    symbols and nomenclature for,
        159–160
mechanical models of School of
        Light physicists of, 157
music analogy of, 183–184
notation system for, 182–183
overlapping, 40–41
patterns in, 178
PhDs in, 80t
quantitative formulas for, 62–63
raie ultime, 179
red cadmium, 190
resonance line, 179
from Saturn's rings, 105–106
sodium, 127–128
solar
    absorption lines of, 20n
    data from Mount Wilson Ob-
        servatory, 195
    standard reference line, 118–119
    stellar, 193 (*See also* specific stars)
        George Ellery Hale's interests
            in, 200–201
        high dispersion studies of,
            204–208
    terrestrial, 57–58
        HAR's study of, 74
        John Mohler's studies of, 75
        School of Light and, 148–152
        solar *vs.,* 102–103
titanium, 205, 208
ultraviolet, 122, 125, 126, 132
    Theodore Lyman's publications
        on, 122
vibrating model of, 156
from water-vapor lines in Mars spec-
        trum, 206–207
St. John, Charles, 159
    revision of HAR's solar wavelength
        tables by, 190–192, 196–198
Stallo, J.B., xxi
Stanley, William, xix

Stark effect, 149–150, 170–171
starlight, spectral lines in, 32
star(s)
    Alpha Orionis, 199
    Arcturus, 105, 108, 148, 154, 200, 201,
        206
    Capella, 199
Steinheil plates, 42
stellar radial velocity studies, 206
Stoletow, A.G., 48
Strong, John, 143, 144–145, 146
Struve, Otto, 48, 208
sun, 18, 22, 57, 197. *See also*
        spectrum(a), celestial, solar
sunspots, 3
surveyor, railroad, HAR as, 4
Sweetnam, George, HAR's work and,
        xxii–xxv
Sylvester, James Joseph, 82–83, 116
symbols and nomenclature for spectral
        lines, 159–160

T

Tacchini, P., 95
telescope(s), 148–149
    200-inch at Mount Palomar, 151–152,
        199–200, 209
    100-inch (Mount Wilson Observa-
        tory), 152
        development by Paul Merrill,
            201–203
        high-dispersion investigations
            performed on, 206–208
    observatories of George Ellery Hale
        and, 91–93, 98
Tesla, Nikola, xix
Thaw, WIlliam, 42
"The Moral Influence of True Science"
        (sermon), 66–67
theory(ies)
    atomic theory
        described as "a grand piano"
            by HAR, 168
        HAR's attitude toward, 114–116
        Niels Bohr's views on, 125
        William Meggers' spectro-
            scopic studies and,
            130–131, 175
    Bohr's theory, 179–180

elastic-solid theory of light, 25
electromagnetic theory of light, 25, 139
grating (of HAR), 116
quantum theory, 19, 180
   of spectral line emission, 178
of relativity, 19, 25, 155–156, 192
wave theory of light, 153
The Physical Review, 97, 126
   nature of spectroscopic articles in, 141–142
   School of Light students' publications in, 79
The Physicists (Daniel Kevles), xx–xxi
The Rise of Modern Physics, 153
Thomson, William, Sir, 48, 138–139, 168
   elastic-solid theory of light, 25
titanium, spectrum of, 205, 208
Trowbridge, John, xviii, 47, 48
tuning fork, 157
Turner, Frederick Jackson, 27n

U

ultraviolet spectra, 125, 126, 132
United States Coast and Geodetic Survey, 65n, 117, 118
University College, London, 9
University of Berlin, 64
University of Chicago, xviii, 91, 131, 135
   George Ellery Hale at, 92
University of Pennsylvania, 11
   HAR's possible employment at, 4
Unsöld, Albrecht, 197

V

vacuum tube, 128–129
Vogel, H.C., 95, 160
von Fraunhofer, Joseph, 13, 21–22, 152
   foundation of diffraction spectroscopy by, 31–34
   solar spectrum observed by, 18
von Helmholtz, Hermann, xviii, 7–8, 48, 102
von Laue, Max, 174

W

Wadsworth, Frank, 148
Wadsworth mounting, 148
Warner, Deborah Jean, 35
water-motor, grating powered by, 38
water vapor, 128–129
   lines, in Mars spectrum, 206–207
wavelength system, international, 190
Weart, Spencer, 17
Webber, George N. (Rev.), 66–67
Webster, Arthur Gordon, xvii
Wiedemann, G.H., 48
Winnik, Herbert, comment: HAR and charisma, 70
Wollaston, William, 18n
Wood, Harry O., 198–199
Wood, Robert Williams, 30–31, 126, 131, 135–136
   acoustical analogy with diffraction gratings of, 30–31
   blazed grating for 100-inch Mount Wilson telescope of, 203–205, 209
   Draper Medal for, 207–208
   Edwin Kemble and, 181
   glass gratings made by, 144
   hydrogen spectral studies of, 128–129
   international ties of, 158
   iodine spectrum studies of 165-166, 165
   mechanical skills of, 23–24
   need for personal recognition of, 208–209
   ruling engine developed by, 143–144
   School of Light recruit, 77–78
   sodium studies of, 127–128
   spectrum creation and, 30–31
   use of "resonance" vs William Meggers, 179
   vs Joseph Ames, 155
Wood, Robert Williams (father), 23
World War I, 126, 128, 150, 158, 175, 176, 201
Wright, Helen, comment: George Ellery Hale, 87–88

X

x-rays, crystals as diffraction instruments of, 174

Y

Yerkes, Charles, 92, 94
Yerkes Observatory, 90, 149, 208
  physical laboratory at, 92–93
Young, Charles, 91, 95
Young, Charles A., 83n

Z

Zeeman effect, 150, 170–171, 172, 194
Zwicky, Fritz, 208–209